T0227873

A Sound Person's Guide to Video

A Sound Person's Guide to Video

David Mellor

Routledge
Taylor & Francis Group

NEW YORK AND LONDON

First published 2000

This edition published 2013
by Focal Press
70 Blanchard Road, Suite 402, Burlington, MA01803

Simultaneously published in the UK
by Focal Press
2 Park Square, Milton Park, Abingdon, Oxon OXI 4 4RN

Focal Press is an imprint of the Taylor & Francis Group, an informa business

British Library Cataloguing in Publication Data
A catalogue record for this book is available from the British Library

Library of Congress Cataloging in Publication Data
A catalog record for this book is available from the Library of Congress

Composition by Genesis Typesetting, Rochester, Kent

ISBN 13: 978-0-2405-1595-3 (pbk)
ISBN 13: 978-1-1384-6882-5 (hbk)

Contents

Foreword   *ix*

1   The origins of television and video   1
In the beginning, Scanning, The coming of colour, A new standard, Video

2   The magic of television   10
Colour television

3   Video recording – the impossible dream   18
Early developments, Helical scan, U-Matic, Domestic video formats

4   The electronic eye   27
Part 1 Camera basics: Tube cameras, Charge-coupled devices, The colour camera.
Part 2 The modern camera: Hyper HAD, Into digits, Master Set-up Unit

5   ENG and Betacam SP   44
Betacam, Compressed time division multiplexed system, Audio in Betacam, The future

6   Digital video   51
D1, D1 error protection, D2, D3 and D5, Why so many formats? DV and DVCPRO: Small is beautiful, Nuts and bolts, DVCPRO, Digital cinematography, Step up from SP, Compression, Yet another format?, The kit, The accessories

7   Standards conversion   73
Three problems, Sampling, Composite video, Interpolation, Motion compensation, Comparing standards converters

8   The video monitor   81
Sync and scan, Antenna to CRT, Display technology, The shadow mask, Progressive scan, Flat panel displays, LCD, Plasma displays, Future technologies

9   Home cinema   96
The vision, Audio, Virtual Surround

10   Nonlinear editing   105
Offline/online, Timeline, Integration

11   JPEG and MPEG2 image compression   113
JPEG, Discrete cosine function, Entropy coding, Results, JPEG for moving pictures, MPEG2, Syntax and semantics, Spatial and temporal compression, Motion estimation prediction, Profiles and levels, Applications

12   Digital television   132
Digital television in the UK: Digital video, Broadcasting, Possibilities, On demand, The consumer angle, Widescreen, Conclusion. Digital television in the USA: Technical issues

13   Film   146
A brief history of film, Meanwhile in France . . ., The coming of sound, Widescreen, Sideways look, Colour, 3D, Why film will prosper

14   Film stock, film laboratories   163
Formats, Types, Intermediate and print film, Laboratories, Rushes, Printing, Editing and regrading, Release print

15   Cinema technology   183
Lamphouse, Reels and platters, Lenses, Sound head, Cinema systems

16   IMAX   192
The camera, Projection, The IMAX cinema, Post-production

17   Telecine   201
Film feats, Technology, Cathode ray tube, Scanning, Digital processing, High resolution

18   Pulldown   209
29.97, PAL pulldown, So what should you do? A new frame rate?

19   Lighting technology   218
Film and video lighting, Lamps glow, bulbs grow, HMI, Broads, Blondes and Redheads, Moving light, changing colour – performance lighting, To boldly gobo

20   The art of bluescreen   234
Rotoscoping, Ultimatte, Motion control

Appendix 1   The science of colour   243
Light, The eye, Subtractive colour mixing, Additive colour mixing, Colour triangle, Colour temperature

Appendix 2   Timecode: the link between sight and sound   251
The nature of timecode, Types of timecode, Timecode generation. Sound and picture, working together: Finger sync, Code-only master, 'Real' instruments, CTL and direction. Synchronizer systems: System extras, Jam on it, Synchronizers, Synchronization terminology

Appendix 3   Audio in video editing   273
The editing process, The end of offline?

Index   281

Foreword

As an audio specialist, you obviously get immense enjoyment and satisfaction from interplay between the technology and artistry involved in the creation of sound images, to inform and entertain your audience. You wouldn't want to work anywhere else, because if you did then you would never have had the required determination to make it into the sound industry in the first place. But sound no longer exists in isolation from other media and art forms. In particular, sound is now so closely bonded to the visual image that we must at all times consider the effect the actions we take will have on the finished production as a whole, in the mind of the moviegoer or TV viewer. *A Sound Person's Guide to Video* covers the technology not just of video but also film, multimedia and live performance as well – in fact anywhere that visual images and sound go together. Some of the chapters in this book will be cutting edge technology, others will look at the history and background to modern techniques. Occasionally these pages will cover aspects of the film, video and related industries themselves to see how the pieces fit together. It's a whole new world out there, so let's take a peek . . .

CHAPTER 1

The origins of television and video

In the beginning

Contrary to the body of popular general knowledge that goes to make up the questions on Trivial Pursuit cards, there was no one person who can be said to have invented television. One particular person, John Logie Baird, was the first to get it to work and make the world aware of what he had done, but he was building on the important achievements of others, without which probably he would hardly have been remembered at all. It was a very logical progression after the invention of radio to think about how nice it would be to transmit pictures over the airwaves, or even along a length of wire. But electronics had not progressed to a sufficiently sophisticated level to make this possible. The cathode ray tube, around which most television receivers and video monitors are based, was first developed by William Crookes in the 1860s, and that important device the triode vacuum tube was invented by Lee De Forest in 1906. The main stumbling block was a detector for the television camera, and if one of these could have been time-warped back to the beginning of the century we would have had a satisfactory television system in operation very much sooner. Because of the impracticality of electronic television in the early days, the only other option was to do it mechanically, and this was developed into a system that was actually used for early experimental television broadcasts.

Scanning

It is no problem to convert a sound signal to electricity and pass it along a wire because one microphone can pick up all of the sound present at a particular point in space, and that is good enough for our ears to recognize the reconstructed sound emanating from a loudspeaker as being a passable imitation of the original. With pictures, the situation is more complex because a picture is two-dimensional, with height and width. It is not possible to transmit this picture along a wire all in one piece because the wire effectively only has a single dimension – length. What is needed is to split the picture up into components and send them down the wire one after the other and then reassemble them at the other end. For instance, in Figure 1.1, you could tell a friend over the telephone which squares were light and which were dark and if you were both working on the same grid they could assemble the same picture. All you would have to do is call out light-dark-light-dark, etc. and periodically say that you had reached the end of a line and that you were starting a new one. If you were transmitting a moving picture in this way then you would also have to say when you had finished one frame and were about to start on the next.

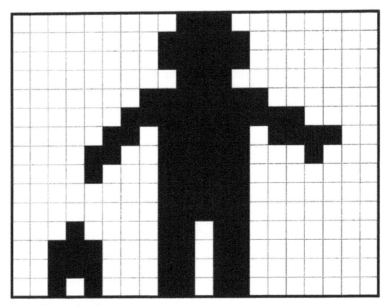

Figure 1.1 Example of scanning.

The first device that could do this automatically was invented by Paul Nipkow in Germany in 1884 and was called the Nipkow disc. Figure 1.2 shows how simple it was. It could scan a scene in a very similar way to a modern camera. The scene would have to be very brightly illuminated and an image focused onto the disc. The disc was rotated and light from only one hole at a time was allowed to fall upon the photocell. This scans the scene into a number of lines, one per hole, and the varying brightness of the scene as the scan progresses causes the output voltage to vary in an exactly similar way. When the disc has made one complete rotation, that completes a frame and a new scan starts. This can also be done in reverse with light being projected onto the scene through the holes.

Light source

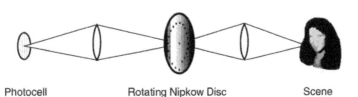

Photocell   Rotating Nipkow Disc   Scene

Figure 1.2 Nipkow disc.

The only problem Paul Nipkow had with his disc is that he never got around to making it. It remained an interesting idea until John Logie Baird started work on his system, using the Nipkow disc scanner, in 1923. His first system was not very sophisticated and had only eight lines, resulting in a very coarse-grained picture. But it worked and it worked well enough for Baird to have to think of a name for the company he wanted to set up to exploit his invention. He called the company Television Ltd. By 1929, Baird had developed his invention to a point where the BBC had become interested and begun regular experimental broadcasts. The number of lines had increased to thirty, but the frame rate was only 12 Hz, which resulted in a very noticeable flickering of the image, although it did have the advantage that the bandwidth was low enough for transmission over a normal sound channel.

The development of electronic scanning was the last piece in the jigsaw that made electronic television systems a practicality. A collaboration between EMI and the Marconi Wireless Telegraph company produced a system that was capable of 405 lines at a frame rate of 25 frames per second. Baird, too, had been busy and had a mechanical system with 240 lines at 25 frames per second, and also a system which involved filming the subject and quickly developing that film before scanning to produce almost live pictures. Apparently this last system was just a little bit unreliable! Known far and wide for their fairness and impartiality, the BBC implemented the Baird and EMI systems and began broadcasting. Not surprisingly, EMI's electronic system was found to be so superior to Baird's systems that the tests, originally planned for two years, were cut short after three months. The 405 line, 25 frames per second standard lasted a long time up to the 1960s when the BBC introduced their new channel, BBC2, on the new 625 line standard only. This meant that to receive the new channel, you had to get a new set, but 405 line broadcasting continued into the 1980s by which time, presumably, all the old 405 line sets – or their owners – had worn out.

Meanwhile in the USA, an engineer called Vladimir Zworykin was inventing a device known as the iconoscope – the forerunner of today's camera tube. This was working by 1932 and although there were intermediate developments, it was still in use up until 1954 when the vidicon tube came into being. The development of electronic scanning was the last piece in the jigsaw that made electronic television systems a practicality. In 1929 the FCC (Federal Communications Commission) licensed a number of stations to make experimental mechanical television broadcasts but over the next few years it became apparent that mechanical TV was not the way to go. In 1935 David Sarnoff of RCA allocated a budget of $1 million to develop a complete electronic television system – an extraordinarily large sum of money during the economic depression. Tests commenced using an iconoscope-based camera using 343 lines at 30 frames per second. 1939 saw an increase in the number of lines to 441 and a more sensitive camera. RCA's system was apparently workable but the FCC were torn between their role in promoting new technology and in controlling the giant near-monopoly of RCA. They determined that full commercial broadcasting would not be allowed until a standard was agreed by the whole of the industry, including Dumont and Philco who had devised rival systems. To resolve this problem a committee representative of the industry as a whole was established to report to the FCC on a system suitable for television broadcast. That committee was called the National Television Standards Committee, or NTSC, and devised the set of standards including the 525 lines and 30 frames per second that is now familiar. This was accepted by the FCC in 1941 and commercial television broadcasting was ready to begin.

The coming of colour

Detailed explanations of how colour television works can wait until later, but in principle it is necessary to transmit three pictures, one for each of the three primary colours red, green and blue. However this is achieved, there are three major problems. The first is that you can only fit so many transmission channels into a given amount of airspace, or bandwidth to use the correct term. If three full-bandwidth pictures were to be transmitted then this would obviously reduce the maximum number of channels available to a third. The other twin problems are of compatibility. In the early stages of colour television, at least, it was very important that it was possible to receive a colour television transmission on a monochrome set, and also that a colour set would be able to receive a monochrome transmission.

Colour television was seen as early as 1928 when Baird devised a Nipkow disc with three sets of holes for the three primary colours, but the first serious proposal for a colour system suitable for broadcasting came from CBS in 1940. To overcome the difficulty of bandwidth, the system was compromised in three ways from the existing monochrome standard of 525 lines at 30 fps and a video bandwidth of 4 Megahertz: the bandwidth was increased from 4 MHz to 5 MHz; the frame rate was reduced to 20 Hz; and the number of lines was reduced to 343. This gave a less detailed picture which would have had a noticeable flicker, but it had the all-important ingredient of colour.

The CBS system was known as a 'field-sequential' system in which the first frame of the programme included red and green information, the second frame carried blue and red, the third green and blue, etc. The problem of flicker might have been bad enough but this was compounded by the problem that a white object moving across the screen would break up into a sequence of coloured images. People wanted colour television, but not that badly! The system also used a mechanical disc to filter the images, which seemed like a nasty hangover from the days of Nipkow discs. Nevertheless, the CBS system was remarkable in that it was television in colour, and that in itself was a significant achievement. They were able to improve the system and by 1949 it had 405 lines at 25 fps and would still fit in a standard broadcast channel.

RCA, perhaps mindful of the mechanical versus electronic debate in earlier days, had decided to wait until it could produce a fully electronic system and was not ready to compete. But CBS had a system and RCA didn't, yet. The FCC were not in the business of blocking working systems at the behest of a powerful rival company because they said that they could do better given time. The CBS colour television system was approved for broadcasting in the USA in 1951, after a decade of debate and arguments that went all the way up to the Supreme Court.

The introduction of colour was a disaster. In the time between CBS developing its system and actually being given permission to broadcast, the number of black and white sets had increased enormously, and because of its incompatibility with these sets the new colour service could only be slotted in at a time when most people were doing things other than watching TV. CBS was legally committed to manufacture sets which no-one wanted and faced severe financial problems. It is also worth mentioning that NBC did not have any particular interest in broadcasting CBS colour!

Fortunately for CBS, the Korean War intervened and the manufacture of colour TV sets was prohibited by the National Production Authority so they didn't have to carry on flogging their dead horse for too long.

A new standard

While CBS had been busy working on a set of compromises that would allow their field-sequential system to work satisfactorily (yet still not give the public what they wanted), RCA had made the important discovery that the eye was not as sensitive to detail in colour as it was to detail in a monochrome picture. RCA proposed that fine detail should be extracted from the red and blue components of the picture and added to the green signal. This would allow the bandwidth of the red and blue signals to be reduced. This developed into a system where monochrome information is transmitted as a full bandwidth signal and the colour as a pair of reduced-bandwidth signals called I and Q. The Q signal represents blue shades, in which the eye is particularly insensitive to fine detail, so that the bandwidth in this signal can be reduced still further. The I and Q signals are modulated onto a carrier in such a way that the colour information slots neatly into gaps that exist in the monochrome waveform and compatibility was maintained with existing monochrome sets. Consisting mainly of RCA's technology, yet with important contributions from others, the NTSC finalized and approved what came to be known as the 'NTSC system' on July 21, 1953. It was accepted by the FCC in December the same year and commercial colour broadcasts were authorized from January 22, 1954.

Shortly afterwards in Europe, politicians were thinking that if they allowed colour television, they would have to use a different system or Europe would be flooded by imports. There was also the advantage that whatever problems had been overlooked in the NTSC system could be corrected in a new European system. The one disadvantage of the NTSC system is that the colour signals are transmitted on the same frequency and are separated only by their phase. This means that if there is any phase problem in the transmission path, the overall colour bias of the picture will change. This led to the NTSC system's description as being 'Never The Same Colour'! The European PAL system, developed by Telefunken in Germany, corrects this problem by alternating the phase of the colour carriers every line, hence Phase Alternate Line. In PAL, any phase errors tend to average out rather than change the overall colour of the picture.

SECAM stands for Sequential Colour with Memory, but translated into French, since it is a French system. Their reasons for having a different system to everyone else can only be guessed at, but it is hardly surprising that SECAM was adopted by Eastern Europe so that if people did tune in to Western TV broadcasts, with all the consumer temptations they offer, they would only receive black and white pictures. Whereas NTSC and PAL are very similar, SECAM operates in a very different way by transmitting only half of the colour information on each line of the picture. It holds that information in a delay line until the next line starts, when it is mixed with the other half of the colour, which is then delayed until the next line, and so on. This avoids having to transmit two colour signals at the same time, but raises complications in signal processing such as special effects.

Video

The difficulty of recording a video picture onto tape was long held to be on a par with finding the secret of eternal youth. RCA's video recorder was developed by Harry F. Olsen who was thinking along the lines of an audio recorder, but able to record a signal with a bandwidth of 4 Megahertz, rather than the tiny-in-comparison audio bandwidth – a mere 20 kHz. His machine used five tracks to record red, green, blue, brightness and sync signals and, after a period of development, he managed to reduce the tape speed to twenty feet per second. Apparently the spools were so large, and their inertia so great, that the engineers were issued with leather gloves in case the brakes failed and they had to slow down the machine by hand! The problems of longitudinal recording were insuperable and the machine never made it into broadcast use. While other companies, including the BBC with their VERA (Video Electronic Recording Apparatus), continued with their impractical efforts, it was a much smaller company that developed the video recorder into a usable device – Ampex.

Ampex was founded in 1944 by Alexander M. Poniatoff, who gave his initials to the company name, plus 'EX', standing for excellence. One could say he was being a bit smug, considering that he started with six employees in a garage, but he showed the mighty RCA corporation more than a thing or two in 1956 when he demonstrated a video recorder that knocked the spots off the RCA demonstration machine. Charles Ginsberg was the head of a development team that included Ray Dolby of noise reduction fame. The idea that made video recording a practicality was transverse scanning. A rotary head is used to write a track from one edge of the tape to the other, which breaks up the continuity of the recording and so would be totally unsuitable for analogue audio, but which is entirely appropriate to the line structure of a video image. Four heads were mounted on a horizontal drum which laid down slightly slanted parallel tracks. The writing speed was about 40 metres per second which gave a response up to 15 MHz. This response above and beyond the 4 MHz bandwidth of a colour TV picture made it possible to use Frequency Modulation (FM) as the recording system which removed the problems of stability and the fact that a TV picture covers a frequency range of around 18 octaves, which was virtually impossible to record onto the tape directly. The Ampex video recording system was called the Quadruplex system, and many Quadruplex machines are still in active use today, even if only as archival playback machines.

The Ampex Quadruplex video recorder was demonstrated at the NAB convention in 1956 and deliveries were made to broadcasters in 1957. Ampex had the field to themselves until 1959 when RCA clawed their way up to a position where they could compete. History can be an instructive subject sometimes because Ampex were so pleased with their efforts that they allowed RCA to sneak up on them in 1961 when they released a video recorder with an all-transistor design, to which Ampex had no reply until 1965 with their high bandwidth Quadruplex recorder.

Although Quadruplex was a great innovation, like all innovations it had its own time, and that ended in the late 1970s. Although the performance could be excellent, it was maintenance-intensive, particularly in keeping the outputs of the four heads consistent to avoid a 'banding' effect. The successor to Quadruplex used helical scanning, which was also an Ampex development, where the tape is wrapped around a drum carrying the head, or heads, and tracks are recorded at an angle across the width of the tape. The main advantage of helical scanning is simplicity, but rather than the one universal format, there came to be a number of competing formats, which brought in other complications.

Helical scanning is used for video recorders from domestic VHS right up to the broadcast digital formats, and in digital audio recorders too. The dominance of Ampex receded in the broadcast video market as they did battle with Sony with the 1 inch helical scan C-Format officially adopted by SMPTE in December 1977. Unfortunately for RCA, they did not

Figure 1.3 The production version of the first practical VTR, the Ampex VR-1000, went on air on CBS November 30, 1956 with a West Coast time delay broadcast of 'Douglas Edwards and the News'. (Courtesy of Ampex Corporation. All rights reserved.)

recognize the advantages of the new format, particularly its reduced maintenance requirement. When it became apparent to them that Quadruplex was on the way out, they were well behind in helical scan development and decided to attempt to go one stage further and develop a digital video recorder. A re-badging deal with Sony on C-Format machines was followed by their own terminally unsuccessful TR-800 which, together with the expenditure on digital research, led to financial losses and the closure of the Broadcast Equipment Division in 1985. A sad end to RCA's involvement in broadcast equipment but their achievements were worthy and are appreciated.

CHAPTER 2

The magic of television

Before television was even a twinkle in John Logie Baird's and Philo T. Farnsworth's eyes, let alone the viewer's living room, other great inventors were grappling with a closely related problem – how to make pictures move. Like television, there isn't one person who can truly be called the inventor of moving pictures, but there are a number of contributors who each brought the state of the art a little bit closer to perfection. Most people now know the basics of how film cameras and projectors work, but back in the days when nobody had the remotest idea that it might be possible to record moving images it took real genius to work out how to do it. Since the whole process relies on certain characteristics of the human eye, and television and video do too, I shall spend a little time on film before moving on to the electronic alternative.

As you know, a moving image is recorded on film as a sequence of still pictures projected onto a screen at a rate of twenty-four per second. The frame rate is fixed by the ability of the eye to 'join up' the images into something that looks like real motion. If the eye/brain combination couldn't do this we would have been in trouble right from the start, since as far as I know no-one has ever produced a moving image in any other way. At a rate of less than 24 frames per second, actions are noticeably jerky and the chosen frame rate is just enough to overcome this. But even at this frame rate, the image on the screen still flickers noticeably. To avoid the flickering, the frame rate could be increased but this would push up film stock costs. A better way to do it is to have a rotating shutter which blanks the screen between frames and once more during each frame, increasing the flicker rate to 48 per second. The eye's persistence of vision holds each repetition of the image long enough until the next one comes along and so the picture looks steady. In fact, a flicker rate of 48 times per second isn't totally adequate and flicker can be noticeable in bright images, especially from the corners of the eye. Computer screens use a flicker rate of 75 Hz or more to produce an image which is comfortable to look at over long periods.

In film, live action is broken down into a sequence of still pictures so that it can be captured and then projected. The television process breaks down the image still further so that it can be sent along a wire, or broadcast. In the camera the image is scanned by an electron beam which traces out a path which starts at the top of the image and works down to the bottom so that the brightness of each tiny segment of the picture can be converted to a voltage and sent down the wire. Figure 2.1 shows the electron beam starting at the top left and scanning the first line of the image; at the end of the line it flies back quickly so that it can start on the next. When it has covered the entire image, it starts all over again. The electron beam in the video monitor or television receiver traces out exactly the same path, controlled by synchronization signals, and reproduces as closely as possible the levels of brightness of the original image on a phosphor-coated screen. The pattern that the electron beam traces out is called a raster.

Just like film, it takes a certain number of complete images per second to give the illusion of smooth motion. Since the mains frequency in Europe is 50 Hz (60 Hz in North America), it was easiest to use this as a reference and use half the mains frequency, 25 Hz (30 Hz in North America), as the frame rate rather than the 24 frames per second of film. But in the same way as a film projector needs a shutter to increase the flicker rate, the flicker rate of TV has to be increased also. This is done by scanning at half the vertical resolution, leaving a gap in between adjacent lines, then going back to fill in the lines that were missed out. This means that half the picture is transmitted, then the other half. The eye's persistence of vision marries everything together and completes the

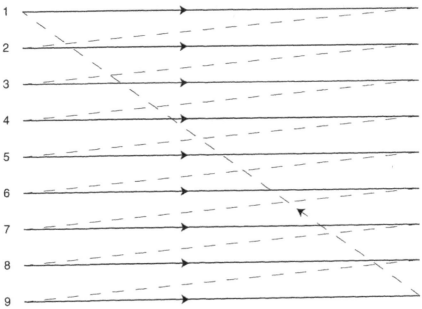

Figure 2.1 Non-interlaced or progressive scan, as used in computer monitors.

illusion. To throw in a couple of technical terms here, each picture half is called a field. The complete picture is a frame. Figure 2.2 shows an interlaced scan. It is theoretically possible to use higher ratios of interlacing than the 2:1 ratio that we do, but the individual lines will start to flicker noticeably and moving vertical lines will appear to fragment, spoiling the illusion.

Now that we have looked at the reasons for the 25 Hz frame rate, let us examine a few other characteristics of the television picture as we know it. In my diagrams I have cut down on the number of lines that go to make up the complete picture – 625 (525 in North America). But did we arrive at the figure of 625 lines by pulling it out of a hat? Of course not! We need to have a certain number of lines to give an acceptable level of definition in the picture, and high definition systems are now in use which have over a thousand lines. There has to be a compromise in the number of lines because the more you have, the more bandwidth has to be used to accommodate them. Broadcasting bandwidth (which is the difference between the highest and lowest frequency components of a television or radio broadcast, plus a safety margin) is a very precious commodity because there is only so much of it, and it has to be carefully rationed out to those who want to use it. The wider the bandwidth of the signal, the fewer different channels that can be transmitted.

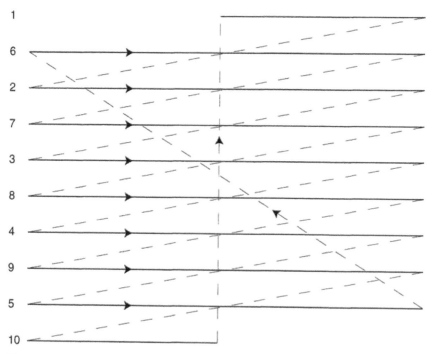

Figure 2.2 Interlaced scanning.

Although the number of lines is specified as 625, there are fewer that go to make up the actual picture. When the electron beam reaches the bottom of the screen after the first field, it has to jump back to the top to start the second. A sequence of broad vertical sync pulses, plus equalizing pulses, occupies the vertical blanking interval between fields when the beam is switched off and repositioned for the start of the next scan. A total of 50 lines per frame are lost in this process, bringing the number of visible lines down to 575.

Vertical resolution is governed by the number of lines, but horizontal resolution is limited by the bandwidth we are prepared to allow for each channel. The faster the spot produced by the electron beam changes from light to dark, the higher the frequency produced and therefore the greater the bandwidth. It would seem sensible to have the same horizontal resolution as vertical and since the distance across the screen is 1.33 times the distance from the top to the bottom, we might expect that the horizontal resolution should be $575 \times 1.33 = 765$ 'lines' (actually individually distinguishable dots rather than lines). In fact, this would be wasteful because it has been found that people will almost always want to sit at least far enough from the receiver so that they cannot see the line structure. This means that the effective vertical resolution is decreased to approximately 400 lines, so the optimum horizontal resolution to aim for is actually 400×1.33 – somewhere around 530 'lines'. This equates to a bandwidth of 5.5 MHz (4.2 MHz in North America), or to put it more simply, the fastest rate at which the spot can change its brightness as it covers the screen is five and a half million times per second. It may sound like a lot, but I don't think anyone will say that our television pictures are by any means perfect, yet.

It is worth taking a look at the signal that carries the video picture along a piece of wire or as modulated radio waves through the ether to our TV sets at home. Unlike an analogue audio signal, which represents sound pressure level varying in time and nothing else, the video signal must carry signals to tell the electron beam how powerful it must be to produce a given degree of brightness, and also the position of the beam at any given time. Figure 2.3 shows the monochrome video waveform, magnified to display just over one line. Unlike audio signals, video signals have set levels which are the same, or at least should be, in both domestic and professional equipment. The full voltage range of the signal is 1 V or 1000 mV, but not all of that is used for picture information. The picture portion of the signal ranges from 0 mV, which represents maximum black, or blanking (actually just a little 'blacker than black'), to 714 mV which is absolute white, or peak white. To tell the receiver when it must start a new line there is a sync pulse. The voltage of this pulse, –286 mV, is always less than that for maximum black so that when the spot rushes back from the right-hand side of the screen to the left, it will reveal no trace of its movements to the viewer. To give the waveform time to stabilize at the end of one line and the beginning of the next there are

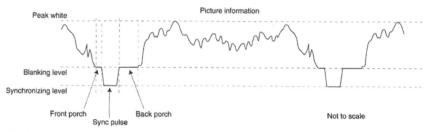

Figure 2.3 Monochrome video waveform.

rest periods known as the front porch and back porch which are nominally 0 mV. At the end of each field is the sequence of field sync pulses that I mentioned earlier.

Colour television

'The quickness of the hand deceives the eye' is the magicians' motto, and it has a close parallel with television and video. If the eye was not so easily fooled, then television – monochrome and colour – would be impossible. If we are to have colour television, then the first thing we must ask is, 'What is colour?' The physical phenomenon of colour is simply the frequency or wavelength of light. Light, as you know, is an electromagnetic wave just like radio waves, but much higher in frequency and shorter in wavelength. There are in theory as many colours as there are wavelengths of light in the visible range – which means an infinite number, of which our eyes can distinguish as many as ten million different shades on a good day. Fortunately the eye uses a very simple mechanism to do this. It has three types of colour sensor: one type is mostly sensitive to red, another mostly to green and the third mostly to blue. These colours, red, green and blue, are called the primary colours for this reason (artists and other people who use pigment colours, rather than coloured lights, use a different system of primaries, which sometimes causes confusion). Obviously, when we see red light it stimulates the red sensitive detectors, and the same for green and blue. When we see the yellow light of a sodium vapour street lamp, which is pure yellow, a colour we have no dedicated detector for, it stimulates the red and the green detectors and we interpret this as yellow. We can fool the eye into thinking it is seeing yellow by showing it a combination of red and green lights in the correct proportion and it will look as yellow as the street lamp. In fact very nearly all the colours that it is possible to have can be simulated accurately by varying combinations of red, green and blue.

The colour systems we use today rely on transmitting a detailed black and white image, which gives the light and shade, on top of which the three primary colours are 'painted'. It is possible to transmit colour by showing the red, green and blue components of an image sequentially, but this has been found to have too many problems and is not as efficient as systems which are better at showing just enough information to fool the eye and no more. At the outset of colour broadcasting (after one false start), it was decided that colour and monochrome broadcasts had to be compatible in both directions – a colour transmission should be receivable on a monochrome set, and a monochrome transmission should be receivable on a colour set. We take this for granted nowadays, of course. So the existing monochrome waveform had to be modified in such a way that TV sets that were already in use would not notice the difference, but the new colour sets would be able to take advantage.

The colour signal is in fact produced as three separate images which then have to be encoded into brightness information, or luminance, and colour information, or chrominance. The red, green and blue signals from the camera have to be combined in a particular way to do this, which once again comes back to the characteristics of the eye. If a white card is illuminated with red, green and blue lights and the brightness of the lights is balanced carefully, the card will appear white. If we measure the brightness of each colour, then out of a total of 100% we would find that the red light contributed 30%, the green light 59% and the blue light 11%. This gives us the equation of colour:

$$Y \text{ (luminance)} = 0.3R + 0.59G + 0.11B$$

Remember that the Y (luminance) signal is the standard black and white information that will be recognized by any set, but has been produced by combining the outputs of a three CCD colour camera in these proportions.

As well as the luminance signal we need to transmit the three colours too, and this is done by combining the three primary colours in the proportions shown:

$$U = 0.49(B - Y)$$
$$V = 0.88(R - Y)$$

You will notice that there are only two chrominance signals, but by combining them with the luminance signal all three primaries can be re-created in the receiver.

As I said earlier, the eye is not particularly sensitive to fine detail in colour, but it is more sensitive to detail in some colours than in others. In North America the I and Q signals, analogous to U and V, are fine-tuned to the eye's response. The I signal corresponds to orange-cyan colours to which the eye is fairly sensitive. The Q signal corresponds to green-purple colours to which the eye is relatively insensitive. To make the best use of this feature of the human eye, the I signal is given a bandwidth of 1.3 MHz and the Q signal a mere 0.4 MHz. Europe's more recent PAL system makes no allowance for this and accords both chrominance signals the same bandwidth.

The bandwidth of the chrominance signals may not seem impressive compared with the luminance bandwidth of 5.5 MHz or 4.2 MHz but it is like making a drawing with pen and ink, and then colouring it in with a broad brush, and it works. The U and V signals are modulated 90 degrees out of phase with each other onto a carrier frequency of 4.43 MHz. This produces a signal with a harmonic spectrum that neatly dovetails into that of the luminance signal, thus exploiting previously unused bandwidth. In North America, the only practical colour carrier frequency unfortunately clashed with the sound carrier frequency. To correct this, the frame rate had to be dropped slightly from 30 fps to 29.97 fps.

The colour information is extracted from the chrominance signal by comparing its timing with a reference signal, the so-called colour burst (shown in Figure 2.4) which synchronizes an oscillator in the receiver. To a moderately close approximation, the phase, or timing, relationship between the chrominance signal and the colour burst is compared to give the hue of the colour. The amplitude of the chrominance signal gives the saturation (how little grey there is). This system is very prone to phase errors between the colour burst and the chrominance signal which

The tennis ball – in or out?

Every match of every tennis tournament the world over suffers from disputed line calls, despite automated assistance. Often TV commentators are reluctant to commit themselves to saying whether they thought the ball was in or out from the slow motion video playback, even when it was clearly one way or the other. There is a good reason for this: if the slow motion playback was derived from a 25 frames per second video recording, you can't tell whether the ball was in or out.

Let us estimate the speed of the ball during play at 50 miles per hour, or 73 feet per second. In the 1/50th of a second between fields (there are two fields per frame, remember) the ball will have travelled almost eighteen inches, so when you see the ball bounce just remember that the 'bounce' didn't necessarily happen when the ball hit the ground, but is simply the lowest point in the ball's travel that was caught by the camera. The real bounce could have happened anything up to nine inches either way, which is more than enough for a ball that appears to be well out of court on the slow motion recording actually to have been in. My opinion is that if a player wants a ball to be judged in, then they should play it well inside the line, just to be sure! Or would that spoil some of the fun?

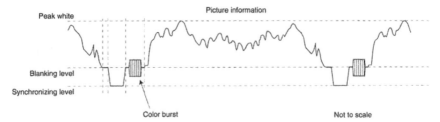

Figure 2.4 Colour video waveform.

obviously will change the hue. This explains the need for a hue adjustment control in NTSC receivers. Some receivers have an auto-tint feature which analyses skin tones. This does have the effect of approximating skin tones to an arbitrary average, and of reducing the range of colours that can be displayed. Auto-tint circuits are best when supplied with an off button! The PAL system corrects any phase errors by causing the hue to shift in opposite directions on alternate lines. The error can then be corrected by simple optical integration in the eye, or electronically.

CHAPTER 3

Video recording – the impossible dream

Imagine, if you can, that the year is 1951. Someone asks you what three technological developments you would like to see in existence in five years time, as birthday presents perhaps. Possible answers might include an all-electronic refrigerator, a light amplifier to make large-scale television projection as bright and vivid as film, and a television picture recorder which can record a video signal just as audio can be recorded on an inexpensive disc or tape and played back in the home.

This was the wish list of David Sarnoff, chairman of the mighty RCA corporation, given as a challenge to his technical staff at a celebratory gathering to mark his forty-five years in the business and the renaming for him of the RCA laboratories at Princeton, New Jersey. To this day, the electronic refrigerator remains uninvented. Large-scale television projectors, such as the Eidophor, which use light valve technology are still rare and have not developed to the stage where they can address the mass market that surely awaits them (conventional projection televisions use very bright cathode ray tubes or liquid crystal light valves and cannot produce a cinema size picture of equivalent brightness to film). Of Sarnoff's three wishes, the only one that has truly come to pass is video recording.

To record a video picture onto tape is nothing short of a technological marvel, and many new devices and processes had to be brought together to make it possible. The early video recorders were bulky and needed careful attention and maintenance, and their picture quality was poor. Now we have camcorders which can almost be hidden in the hand and give truly amazing results. If Sarnoff were alive now he would surely be impressed at how completely at least one of his wishes came true. As for the rest, we shall have to wait a while longer.

Early developments

In the early days, the three major obstacles to successful video recording were bandwidth, timing stability and linearity.

A video signal contains a much greater range of frequency components than an audio signal, which we normally accept as being from 20 Hz to

20 kHz. In complete contrast a video signal contains components from virtually d.c. (0 Hz) to 5.5 MHz. This means that the highest frequency in video is around 200 times the highest frequency you would ask an audio recorder to reproduce properly. If an audio recorder can cope with 20 kHz at 15 inches per second, then a video recorder needs a tape speed of at least 250 feet per second. Needless to say, this is not sensibly possible. Timing stability is vital for a video picture because any jitter will break up the image and make it unsteady, just as wow and flutter in an audio recorder produces pitch variation effects. Also, because colour television relies on having a precise phase relationship with the colour burst reference and the colour carrier, lack of stability will make colour reproduction impossible. The quality of linearity basically means that any changes to the input signal are reproduced exactly by the output. In audio recorders, nonlinearity leads to distortion and added harmonic products. In a video recorder it would mean that the various shades of grey would not be given their correct values, and colour reproduction once again would be badly affected.

One of the earliest prototypes addressed the bandwidth problem by splitting the signal into five components: one for each colour, red, green and blue; one for high frequency components; and one for the sync pulses. After development, the tape speed was brought down to a mere 20 feet per second, but the stability and linearity problems remained. It is notable, however, that from the start, this prototype (from RCA) was capable of handling colour, and by recording the colour information directly it circumvented some of the problems that are caused by lack of timing stability. Unfortunately, this line of attack proved to be a dead end, and indeed at this time no-one knew whether video recording to broadcast standard would ever be possible.

As I outlined in Chapter 1, the great breakthrough was made by a small company called Ampex which subsequently rose to greatness in its field. A six-man team headed by Charles Ginsberg, and including Ray Dolby of noise reduction fame, put together all the necessary technologies over a period of four years' hard work. The bandwidth problem was solved by moving the heads (now four of them) as well as the tape (Figure 3.1).

It seems obvious now that it is the relative motion between head and tape that is important and that a high writing speed can be achieved in this way, but in its day it was a real breakthrough. Moving the heads in itself has a stabilizing effect, but more precise control over jitter was achieved by using an adjustable vacuum tape guide which held the tape at the correct position with respect to the heads. The third problem of linearity was solved by recording the signal using frequency modulation, which side-steps the inherent lack of linearity in the magnetic tape recording medium.

The first broadcast standard video recorder using the Ampex Quadruplex system had at its heart the rotating head wheel with four heads mounted at 90 degree intervals. This wheel spins at 14 440 rpm (in an

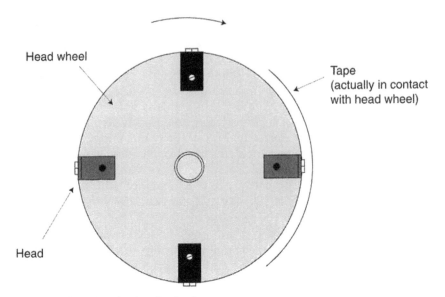

Figure 3.1 Quadruplex head wheel.

NTSC machine) and the heads traverse the 2 inch wide tape at a writing speed of nearly 40 metres per second. Intimate contact between the head and the tape is vital because, especially at high frequencies, any slight gap will cause a momentary drop-out in the signal. To achieve a good contact, the heads protrude slightly (by 50 microns) into the tape, which is held in exactly the right position by a cylindrical tape guide incorporating a vacuum chamber. The actual tape speed from reel to reel is 15 inches per second, which is so slow in comparison to the writing speed that the tracks which contain the video information are written almost at right angles to the tape, at 89.433 degrees in fact. Because of this, the system is known as transverse scan.

The wrap angle of the tape around the head wheel is rather more than 90 degrees so some redundant information is recorded on each pass, but to build up the complete picture either sixteen or seventeen lines are retrieved from each track and stitched together with the next block of lines retrieved by the following head. It takes thirty-two passes across the tape to make up a complete video frame. All the switching that is involved has implications for picture quality. For instance, if the outputs of the heads are not exactly matched there will be noticeable banding across the screen. If the tape is not penetrated by the head by just the right amount, controlled by the vacuum guide, there is a 'Venetian blind' effect which splits the picture into horizontal blocks that are out of alignment. Compatibility between machines was not spectacularly good, due to the tight tolerances involved, and even a small deviation from a true 90 degree head spacing would cause significant distortion of the image.

You'd better have a skilled technician on call if you want to have one of these brutes in your studio.

In addition to the video information area, there is also a control track for sync pulses. The control track acts almost like film sprocket holes to align the transport and head mechanism to the picture. Two audio tracks are available, although one is very narrow and is intended for comments or editing cues. The main audio head is 9 inches away from the video head wheel, which added one more difficulty in the days of cut and splice editing of video tape. Stereo operation was possible on some machines by splitting this audio track into two.

Recording an audio signal onto tape without gross distortion was made possible in the 1940s by the invention of high frequency bias where a tone of 80–100 kHz is added to the audio signal to help magnetize the tape and shift the signal away from the region of nonlinearity. Since the video signal is a much higher frequency than even this, bias is not usable and therefore it is not possible to record a video signal onto tape directly without distortion. Also the frequency range of a video signal is of the order of eighteen octaves, the difference between the highest and lowest frequencies is too great – the output from the playback head would be far too low at low frequencies. The solution to this problem was to use frequency modulation where a carrier frequency is recorded onto the tape, and the frequency of this is altered according to the amplitude of the signal. Although this carrier has to be even higher in frequency than the video signal, the range of frequencies generated is much smaller. The original monochrome Quadruplex system had a video bandwidth of 4.2 MHz, a carrier of 5 MHz modulated between 4.28 MHz and 6.8 MHz from the tip of the sync pulse to peak white with sidebands (extra frequencies generated by the FM process) extending to 11 MHz. The frequency modulation system worked, cured the problems of non-linearity, and is still employed in analogue video recorders today.

The Quadruplex system was gradually improved, as you might expect. RCA added colour by developing a circuit that, since it was impossible to eliminate jitter from the system, would make the colour burst and colour signal jitter along together, thus maintaining the important phase relationship between them. Ampex responded with a High Band version of Quadruplex and RCA, after a struggle, developed a machine to match. Quadruplex machines continued in production until 1981, although by this time they were mainly used for archival purposes. After a long period of refinement, they were capable of a very high quality of picture reproduction, but which suffered badly if the machine was poorly maintained.

Helical scan

The biggest drawback to Quadruplex was the fact that signals from four very accurately aligned heads had to be married together to produce a complete video picture. Helical scan recorders eliminate this problem by slanting the track at a much shallower angle to the tape, making it long enough to contain a complete picture field (half a frame). Since this can be covered by one head it provides inherent continuity, and there is also scope for still frame and slow motion. The first helical scan recorders used 2 inch tape like Quadruplex, but a multiplicity of standards emerged on tapes as narrow as a quarter of an inch. This lack of standardization delayed the acceptance of helical scan by broadcasters for the best part of a decade.

Despite the advantages of helical scan, a further important advantage being lower tape consumption, there are bound to be some problems. One main problem is that the jitter or flutter performance is not as good because the forward motion of the tape now contributes significantly to the combined tape/head motion. Also, since the scan is now much longer, some 16 inches on a C-Format machine, it is difficult to keep the track in a straight line due to uneven tension across the width of the tape. But improved technology has an answer to these difficulties.

There were three main competitors for the prize of developing the standard format for helical scan recording: Ampex, Sony and the German company Bosch. Each manufacturer had of course made a large investment in their chosen system and was unwilling to stand aside and let a competitor impose a standard. Although the Ampex system had already been recognized as SMPTE (the Society of Television and Motion Picture Engineers) Type A, the matter went again before SMPTE and the resulting Type C was a compromise between the Ampex and Sony systems, although probably incorporating a higher percentage of the latter. Bosch's system had already won wide acceptance in Europe, and although it used a segmented scan like the old Quadruplex system, it had its own advantages and was adopted as the SMPTE Type B format. C-Format, however, became the *de facto* world standard of its day for broadcast video recording.

Helical scanning uses a head drum on which one or more heads are mounted. C-Format allows for six heads, although only one video head and one sync head are obligatory. The tape is wrapped very nearly all the way round the drum in an omega shape. Other helical scan recorders may use alpha wrap or half wrap (Figure 3.2).

The other heads in C-Format are two extra sync heads, a playback and an erase head. The extra heads cope better with the gap where the main record/playback head is not in contact with the drum and fill in missing information. The erase head is used for editing where signals need to be placed on the tape with exact precision. C-Format also includes three audio tracks, one of which may be used for timecode, and a control track similar to that in Quadruplex. Not a universal C-Format feature, but an interesting device nonetheless, is the vacuum capstan (which means that the machine needs an air compressor to operate!) which pulls the tape towards it, maintaining firm contact. Also, the tape rests against the

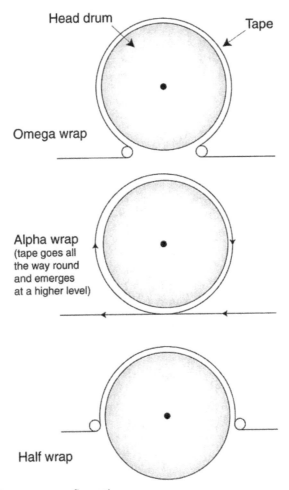

Figure 3.2 Tape wrap configurations.

capstan in all transport modes which means that changing from wind to play is a quick operation. It is possible for a C-Format machine to come up to speed and be synchronized to a studio's master sync generator within four frames, which is impressive compared with early Quad machines which had to be allowed 20 seconds to do the same.

The instability of helical scan compared with Quadruplex is solved by a technique known as timebase correction. A timebase corrector is a device where a jittery video signal can be fed in and all the lines, fields and frames are 'straightened up' to fall into the precise time slots necessary to produce a stable picture. The Ampex VPR3, for example, used a digital timebase corrector in the days before digital video recording was practical, so we have been watching digital images on TV for longer than we think. Timebase correction used to be a big thing in broadcast video and, being expensive, marked a bold dividing line between the broadcast and corporate/industrial video markets. Modern digital video systems by their nature incorporate timebase correction and achieve image stability down to pixel level, where even the best timebase correctors for analogue video could do nothing about timing errors within lines.

U-Matic

Another important development which ran in parallel with C-Format was the U-Matic video cassette recorder, the smaller cousins of which we now see under our TV sets. U-Matic was never intended to be a broadcast medium, but with low tape costs and ease of handling it proved a very useful tool for use in a broadcasting studio for viewing copies and other applications where high picture quality is not necessary. U-Matic uses helical scanning with one 7 inch track holding a complete video field. The writing speed of just over 10 metres per second is well down on the professional formats and accounts for the lower stability, poorer picture quality and greater susceptibility to drop-outs.

The head drum of a U-Matic uses half wrap and therefore two heads mounted at 180 degrees from each other are necessary to maintain a continuous picture. Since the writing speed is low, which therefore curtails high frequencies, it is not possible to modulate the video signal directly onto an FM carrier so the colour information is first stripped out. This technique is known as colour under and is applied to other cassette-based systems too. The colour under technique takes the colour information, reduces its bandwidth and therefore the detail it contains, and positions it at the low frequency (below 1 MHz) end of the video spectrum, from which the luminance (brightness) signal is excluded. The bandwidth of the luminance signal is also restricted. On playback the colour signal is extracted and combined back with the luminance signal so that it once again forms a complete colour waveform that can be recognized by a video monitor.

Domestic video formats

Although we are naturally more interested in what the world of professional video has to offer, sometimes important developments come to light in products intended for the domestic market. One such technique is employed in both the VHS and Betamax formats, although it was first used in a Philips product. All conventional recording on magnetic tape, up until this development, used a guard band to keep separately recorded tracks separate on playback. In audio recording, low frequencies especially tend to 'spread' across the tape and there is always

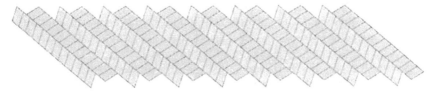

Figure 3.3 Azimuth recording. Adjacent tracks are recorded at different azimuth angles to reduce crosstalk.

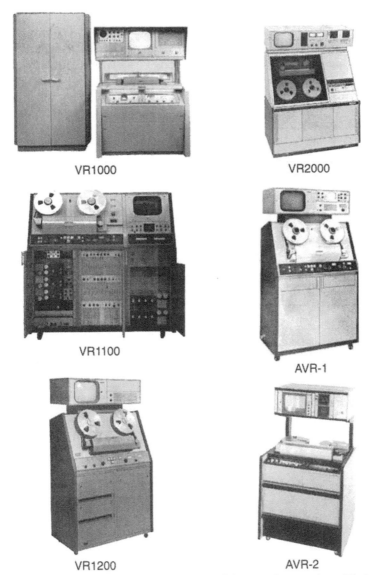

VR1000

VR2000

VR1100

AVR-1

VR1200

AVR-2

Figure 3.4 Ampex family tree. (Courtesy of Ampex Corporation. All rights reserved.)

some crosstalk between left and right channels in a stereo recording. Video recorders used guard bands too, which meant a certain amount of wasted tape. The solution to this wasted tape is of course to eliminate the guard bands and record tracks right next to each other. To deal with the crosstalk, the two heads on the video cassette recorder's head drum are mounted at different azimuth angles, as in Figure 3.3. When the tape is played back, crosstalk is attenuated because the gap of each head is at a different angle to the recordings on adjacent tracks. This technique has also found use in digital audio: in a DAT recorder, the tracks are actually overlapped while recording, making sure that the optimum recording density is obtained. The lessons learned in VHS and Betamax have been extended in the S-VHS and Video-8 (subsequently Hi-8) formats which exploit advances in tape formulations to achieve a better quality image at a cost acceptable to the domestic market.

CHAPTER 4

The electronic eye

PART 1   CAMERA BASICS

It takes a lot of skill to be able to select and position microphones to capture absolutely the best sound for recording or broadcast. To be a boom operator is not perhaps the most glamorous job in sound, but nevertheless it takes a good deal of skill and concentration. Equipping talent with radio mics takes technical expertise and the ability to work with people suffering from pre-performance nerves. Everyone knows how to get a microphone to work – a simple matter of 'plug and point' – but to get the best out of it is entirely another matter. The same applies to video cameras. We all know how to shoot a home video, but we conveniently ignore how shaky and how poorly framed our images are, how we didn't manage to capture the most important part of the action, how the autofocus let us down. Excuses, excuses! TV camera operators are at the sharp end of the broadcasting business and have a difficult combination of technical and artistic skills to acquire. A TV camera is a complex piece of equipment, and it is expensive. A fully specified studio television camera complete with lens, viewfinder and mount could, if you demand the best, easily take you over the £100 000 mark. When you are talking about broadcast video, you are talking big money, and this money buys equipment that is absolutely 100% professional, built to do the job and respond sensitively to the requirements of the operator and director.

Most of us are familiar with video cameras in one form or another. Cameras for domestic use are incredibly inexpensive, thanks to mass production, and offer a quality which would once have been thought to be unattainable. But the difference in picture quality between amateur or semi-professional equipment and cameras for broadcast is remarkable, particularly when you compare them directly on a good quality monitor rather than via a somewhat degraded television image. Let us look at the development of electronic cameras to find out how they work, and what problems had to be solved along the way from primitive beginnings to modern state-of-the-art sophistication.

Tube cameras

In order to transmit a picture along a wire or across the airwaves, the image first has to be manipulated in such a way that it can be transmitted one element at a time. This is done by scanning the scene so that it is split up into a number of horizontal lines, and the different brightness levels can be sent to the receiver, one line at a time, until the entire picture has been built up and another one can commence. The earliest television cameras used mechanical scanning devices. The scene was focused onto a spinning disc drilled with a spiral pattern of holes, through which the light would pass onto a photosensitive cell which produced a varying output voltage proportional to the brightness of each part of the image. This worked, but was inevitably bulky and the high mass and inertia of the components imposed a limit on how far it could be developed. It was recognized very early in the development of television that if it could be possible to scan an image electronically, that would be the way to go. If the scanning device had no physical moving parts, then it could be more efficient and in due course be developed to a high degree of quality.

After mechanical scanners, the obvious way to transmit and receive television pictures was the cathode ray tube. This had been recognized as early as 1908, but the apparent difficulties were so immense that no serious development work was done. Even the originator of the idea commented, 'I think you would have to spend some years of hard work, and then would the result be worth anything financially?' It is impossible to blame him for thinking this way, but we can see now that it is a repetition of the belief centuries earlier that it would be no use having printing presses to mass-produce books because hardly anyone at the time could read!

It is worth having a look at the cathode ray tube in more detail simply because it is such a commonplace device and every home does indeed have one. In a fairly simplified form it looks like Figure 4.1. A glass tube is fitted with two electrodes and most of the air pumped out. The cathode is negatively charged, which supplies it with an excess of electrons, and is heated so that they boil off into an electron 'cloud' around it. The anode is positively charged, which provides an attracting force for these electrons which are now looking for a home. The electrons accelerate to a high speed moving towards the anode – a few may find their mark but most achieve a high enough momentum to pass through the central hole and strike the phosphor screen. The phosphor dissipates the energy of the electrons which collide with it in the form of visible light. This forms a bright spot which can be positioned on the screen by feeding an electric current to the coils which creates a magnetic field that will deflect the electron beam. If the strength of the electron beam can be modulated, this simple device will become a picture tube suitable for a television receiver. And in reverse, with some modifications, it can act as a detector too. That was the difficult part which took a number of years to develop fully.

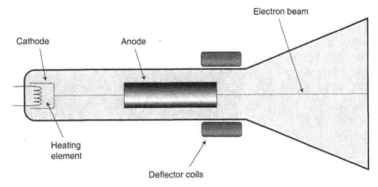

Figure 4.1 Camera tube.

The first fully viable electronic detector was the iconoscope. Modern camera tubes work in a fundamentally similar way and inventor Vladimir Zworykin, who escaped the Russian Revolution to work for Westinghouse as a research engineer, would derive great satisfaction from the fact that even now tube cameras can produce an image comparable with the best any non-film technology can offer. Figure 4.2 shows the iconoscope in cross-section. The electron gun comprises the cathode, anode and scan coils of the simple cathode ray tube described earlier. This fires electrons at the photomosaic onto which the image is focused. The electron beam is scanned across the image in the familiar pattern of horizontal lines – the raster – from the top of the scene to the bottom, sometimes hitting light spots projected onto the photomosaic, sometimes dark according to the varying brightness levels of the scene.

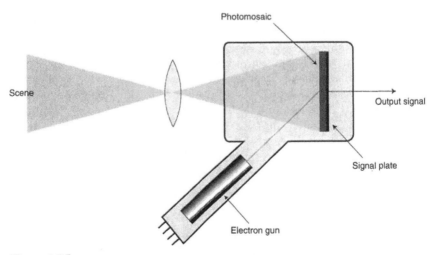

Figure 4.2 Iconoscope.

The photomosaic of the iconoscope is made of a material which emits electrons in the presence of light. Therefore where the scene is bright, more electrons spin off from its surface, which leaves a positively charged area behind. When the electron beam – a stream of negatively charged particles – strikes an area which has become positively charged due to being exposed to a bright part of the image, the electrons are absorbed into the material to balance the charge back to zero. The resulting current is transferred capacitatively to the signal plate from which the output signal is taken in the form of a varying current. With the addition of sync information, a complete video signal is formed.

There were other attempts at this early stage in the development of electronic scanning, such as Philo T. Farnsworth's image dissector. But the iconoscope had the advantage that, although the electron beam could only strike one particular place at any one time, the whole photomosaic was active the whole of the time. Even when the electron beam was active somewhere else, each part of the photomosaic was releasing a greater or lesser number of electrons according to the brightness of the image, ready for when the electron beam came round again to equalize the charge and add another element to the complete picture. Farnsworth had more than a few good ideas in this field, however, and RCA was forced to pay licence fees for the use of his patents. According to legend, the signing of the deal brought tears to the eyes of one of RCA's lawyers!

All camera tubes work in basically the same manner as the original iconoscope, in that the image is focused onto a sensitive target which is scanned by an electron beam. Improvements to the iconoscope were mainly in the design and construction of the target. One could fill a book with all the subtle and ingenious variants of the camera tube and their methods of operation, but I would like to restrict myself to just the next stage of development, which is still understandable to the average non-expert – such as myself – in photoelectric effects and related phenomena.

It's an ill wind that blows nobody any good, and World War II was as ill a wind as they come, but it did promote many advances in technology, for better and for worse. One of the better effects was an improved camera tube, the image orthicon. This was developed for unmanned flying bombs which were to be used to destroy heavily guarded submarines in positions along the French coast. It may come as something of a surprise to learn that the idea of transmitting television pictures from a guided missile was not first used in the Gulf War of the early 1990s but almost fifty years earlier. Apparently the weapons system was not as successful as had been hoped, but the image orthicon was. It was a very precisely designed and constructed piece of equipment which might not otherwise have been developed by purely commercial interests for many years. The scene is focused onto a photocathode which emits electrons from its rear surface to form an electron image on a target mesh. Most of the electrons pass through the mesh and strike a thin glass membrane from which there is a secondary emission of electrons leaving the glass positively charged. The glass membrane absorbs electrons from the scanning beam and a return beam is formed which becomes the video signal, after conversion from being a negative image where current is higher in the darker regions. The advantages of the image orthicon are its higher sensitivity and also the fact that there is some 'overshoot' at the boundaries between light and dark regions, which gives the impression of greater image sharpness.

One of the greatest problems with most types of tube is that they are very sensitive to damage from too much light. This is likely to happen when the camera is accidentally pointed towards the sun, and the cost of tube replacement is not something you would like to bear too often. Tube replacement could cost several thousand pounds, and there is the possibility that the new tube (in a three-tube colour camera) will not match the other two. Even if the camera is pointed at bright lights rather than the sun itself then there may be problems of the light compromising the image or causing point damage to the tube. For instance, a bright window viewed from indoors might superimpose an annoying pattern on images from the camera for a while later. If the light source is bright enough this can become permanent and the tube is irreparably damaged. Damage often takes the form of pinpoint burns which, although they do not make the tube unusable, do cause annoyance.

In addition to this, tubes in general do not have an adequate 'dynamic range' as sound people might call it. They cannot cope with the variations in brightness levels that are typically found in the real world. Studio lighting for tube cameras must be very flat and even compared with the range of brightness levels you would find else-where. I have even seen TV lighting people wandering round with light meters checking the consistency of the illumination, since if the camera tube cannot cope with the scene adequately then there can be no correction later. Other problems include dark current, a signal which comes from the tube even when it isn't illuminated. This leads, in a colour camera which uses three tubes, to another problem: if the dark currents are not matched then areas of shadow in the scene will come out with colours they should not have had. Some types of tube suffer from this particularly badly because the dark current is temperature dependent. Another problem which is very noticeable is sometimes known as 'comet tailing', which occurs when the electron beam does not completely wipe away the image from the target as it scans across. This blurs the motion between frames and degrades the sharpness of moving images. If these problems were not enough, tubes are prone to geometry errors where, for instance, a circle may come out as an ellipse. In a three-tube colour camera these geometry errors lead to problems in the registration of the red, green and blue images, which also occur if the tubes are not in precise alignment.

Charge-coupled devices

The problems of tubes led to the demand for a better type of detector – the CCD. Perhaps in the early days it was not exactly better, but it had fewer problems and could be less expensive to buy and maintain. Charge-coupled devices, or CCDs, have been used for a variety of purposes including audio delay lines. The basic idea is that an electric charge can be passed along a line of capacitors controlled by carefully timed switches. The CCD as found in a video camera consists of an array of silicon photodiodes, hundreds of thousands of them. Each of these acts as a capacitor, the charge on which varies according to the amount of light falling upon it. Each capacitor is connected to the next one along the line by a buffer amplifier and a switch. The switches are left closed as the capacitor collects the light and builds up its charge, then they are all opened simultaneously and the charge level on each element is passed down the line to the next. In this way, the various brightness levels can be 'marched' down the line, collected and assembled into a picture line. The advantages of CCDs over tubes are several: they are much smaller and are more resistant to damage; their geometry is better; registration in colour cameras is not a problem; they are not subject to lag or burn. It is not surprising that they have almost completely taken over from tubes in cameras at all levels from domestic to fully professional. But CCDs are not without problems themselves. One of the most annoying is vertical smear. This occurs when there is a bright light in the scene, where charge will leak along the line of photodiodes creating a vertical bar in the picture. The interline transfer type of CCD will always be prone to vertical smear, but this is only a problem if it is used in difficult lighting conditions. In controlled studio conditions then it will be possible to ensure that excessive contrast in the scene is eliminated and so there will be no vertical smear and the image will be of good quality. If a camera is to be used for outside broadcasts, or in situations where there may be bright lights in the picture area, then it should have a different type of CCD of the frame interline transfer variety which reduces vertical smear to an imperceptible level. Naturally, this type of CCD costs a lot more so it would not be feasible for every camera to be so equipped.

Another problem with CCDs is fixed pattern noise. This was more noticeable in earlier CCD cameras, especially with a lot of shadow area in the picture. With a tube camera, shadow areas would be dark but with a shimmering noise pattern superimposed. This is exactly analogous to the hiss you hear during quiet passages in an analogue audio tape recording. The eye, however, soon gets used to this kind of noise and it can, if not excessive, often be ignored. With CCDs the noise pattern is static because the image is broken down into fixed pixels rather than the continuous scan of the tube. Since each CCD element will have a slightly different sensitivity and is fixed in position, the noise overlays the picture as though drawn onto the inside of the picture tube. Needless to say, this is very disturbing to the eye, but the problem has now been largely solved by developing CCDs with a greater signal-to-noise ratio so that the fixed pattern noise, though still present, is kept to a very low level.

The complete absence of lag in a CCD image means that pictures can be captured without blur. Cameras can be fitted with a variable 'shutter speed' in the same way as a high shutter speed on a still camera can stop fast action. This is particularly valuable if a recording is to be played in slow motion or still frame.

The colour camera

As explained earlier, colour images can be reproduced by transmitting three signals, one for each of the primary colours red, green and blue. To do this, a camera obviously needs to have sensors for each of the colours. In a tube camera, this means having three tubes. Since light enters the camera through only one lens, it has to be split up optically into the three components, which requires a fairly complex system of mirrors or prisms. The prism option is shown in Figure 4.3. The dichroic coatings of the prisms have the property of reflecting one colour and allowing the others through. Light first encounters the blue dichroic surface which reflects the blue component of the scene and transmits the other colour components. The red dichroic coating separates out the red component and allows

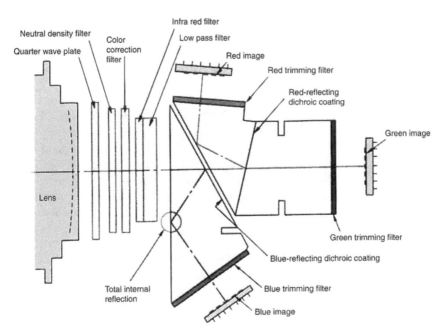

Figure 4.3 Separation of the image into red, green and blue.

what is left, the green, to pass through. Since the reflections in the paths of the red and blue light cause the images to be reversed, additional reflecting surfaces are included to correct this. Even though it is perfectly possible to make a single CCD with detectors for red, green and blue, rendering this complex colour splitting system unnecessary, high quality cameras all use three CCDs, one for each of the primary colours, which leads to highest performance in terms of resolution and signal-to-noise ratio.

PART 2   THE MODERN CAMERA

How did you first become interested in audio? Did you have a little cassette recorder with a built-in microphone that you used to record your parents, siblings, friends, pets and anything else you could find that would make an interesting sound for you? Now of course you have a multi-million dollar studio with a mixing console half a mile wide and racks of expensive equipment reaching almost to the ceiling (or maybe you're still getting there!). At home you have a camcorder which you use for pretty much the same purpose as your old cassette, except now you have pictures (and kids to shoot perhaps). The difference between your home camcorder and a full broadcast quality camera is of the same order of magnitude as the difference between a cassette recorder and your studio, in facilities, cost, and the skill required to work it. Granted, broadcast video equipment has a lot of technology in common with home video, but in terms of features and image quality it is at a vastly higher level. Let us move directly to the heart of any video camera, the CCD.

Hyper HAD

Sony's high end cameras incorporate the Sony Hyper 520 000 pixel HAD 1000 CCD imager, a more highly developed form of the CCD pickup that you would find in a domestic camcorder. Firstly, to explain the terminology: HAD stands for Hole Accumulated Diode where the standard CCD element is enhanced with a layer that accumulates electrical holes (a hole in a semiconducting material is a space where an electron could be, but isn't – if you see what I mean). The advanced electronics section of your local bookshop should be able to provide further technical information, but I can tell you that the end result is a CCD pickup that has a reduced dark current, by a factor of ten. Dark current is the phenomenon where the CCD will still give a signal even in the complete absence of light. Since all the elements of the pickup will have slightly different dark currents the result is 'fixed pattern noise', which is like a still frame of the dancing dots you see on a TV that isn't tuned in to a transmitter, superimposed at a very low level upon the image. Fixed pattern noise used to be a major problem in CCD cameras but with HAD technology it is hardly noticeable at all.

Another feature of Sony's Hyper HAD is the OCL or On-Chip Lens technology. In any Frame Interline Transfer CCD, the boundaries between the individual pixels form the frame storage area which stores image information until it can be transported out of the CCD via the vertical shift register. The frame storage area takes up a certain amount of space which cannot be used to collect light, so much of the light focused on the CCD falls in areas where it cannot be detected. The solution is pretty obvious when you think about it – all you need is a tiny lens over each element and light can be gathered over a wider area. More light equals more sensitivity and less noise, and that is just what you get with a Hyper HAD pickup. In fact the signal-to-noise ratio is typically 62 dB, which is very good in video terms – a little better than the theoretical signal-to-noise ratio of a 10 bit video recorder. In conjunction with OCL technology, Sony have improved the masking of the frame storage area which has the effect of reducing the leakage of light from where it should be to where it shouldn't. In a CCD without OCL, signals can bypass the read-out gates of the elements and break through into the frame storage area carrying the picture information vertically up the chip. The result is vertical smear where a bright vertical bar is seen to intersect bright highlights on screen. Vertical smear in a Hyper HAD CCD is reduced to around −140 dB and is therefore virtually invisible.

Sony claim that the performance of the Hyper HAD 1000 imager is such that it can match or better the performance of a $1\frac{1}{4}$ inch Plumbicon tube which has long been accepted as a benchmark for studio cameras. In terms of dynamic range and accuracy of colour, the two are equal. But Sony reckon the Hyper HAD 1000 is significantly superior in depth of modulation (a more precise way of expression resolution), sensitivity, handling of highlights and the all-important practical issues of size, power consumption, reliability and running costs. Also, although the pixel count of the Hyper HAD 1000 at 520 000 is lower than in some other CCDs, the balance of performance factors is considered to be superior overall, in Sony's opinion, in terms of dynamic range, sensitivity, vertical smear and power consumption.

Into digits

Digital signal processing has reached television cameras and the signals from the three Hyper HAD 1000 CCD imagers is almost immediately converted into digital form which offers greater possibilities for correction and manipulation than analogue signals would. Before the analogue-to-digital converter, however, there are white balance, pre-knee and gain boost processes in the analogue domain. To provide a full range of adjustment of these parameters would require an analogue-to-digital conversion resolution of more than 13 bits, which would be overkill as far as anything further down the line is concerned. Modern digital video recorders, for instance, resolve to 10 bits, and even this is considered by many to be more than adequate. White balance will be familiar from home video where the camera must be adjusted to the colour temperature of the light under which the images are being shot. Light from filament bulbs is much redder than daylight, and although our eyes don't seem to notice, the camera certainly does. White balancing is done automatically these days simply by pointing the camera at a white card and allowing the camera to calculate the corrections necessary (sometimes it is simply judged by eye). Pre-knee reduces the dynamic range of the CCD from 600% down to 225% suitable for 10-bit conversion. '600%' refers to the 100% of peak white – the brightest highlight your TV can display. The range beyond this can optionally be brought down in level to enhance highlight detail and give more of a 'film' look. Gain boost is used where a scene is very dark and is still not bright enough even when the lens is opened up to its widest aperture. Plainly if a scene is too dark after analogue-to-digital conversion, then boosting the level will raise the noise floor. A better result is achieved if gain is applied in the analogue domain before conversion.

I have said that the resolution of the analogue-to-digital conversion process is 10 bits – rather less than the 16 we are used to in audio. But when it comes to sampling rate, video is streets ahead. You can forget about the 44.1 or 48 kHz sampling rates of audio – a digital video camera samples at 18 MHz! It makes me wonder why we argue whether it is worthwhile doubling the audio sampling rate to 88.2 kHz or 96 kHz. These figures are so tiny in comparison to what is already being done in video, why on earth are we restraining ourselves when the technology is already available?

In the digital domain, one of the most important DSP functions is detail correction. Fine detail in an image is given by high frequency components in the signal. In audio of course we might seek to brighten a signal by boosting the high frequency content, or add a controlled amount of distortion using an exciter. Similar processes apply to video, too, where sometimes we would want to see the maximum amount of detail, and even emphasize it, whereas on other occasions too much fine detail can be a problem – wrinkle concealment is not just a matter of makeup these days. Fine detail in an image is enhanced, not by boosting the level of high frequencies which would simply make them brighter, but by increasing the contrast at transitions between light and dark. Figure 4.4 shows how a boost is applied at an edge, which makes the transition just that little bit better defined. A high end camera would be blessed with 'variable horizontal detail peak frequency' with which the width of the detail correction signal can be adjusted to compensate for different shooting conditions and lens settings. It would of course be possible to apply too much detail correction to some subjects so that side-effects

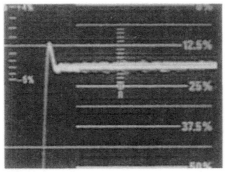

Figure 4.4 Detail enhancement.

appear, which include 'black halo' and 'stepped diagonals'. Black halo is where thick black edges surround bright objects; stepped diagonals are simply jagged slanted edges, which you have almost certainly encountered. These problems are reduced by clipping the detail correction, taking account of both horizontal and vertical directions. There is of course such a thing as too much detail enhancement, particularly of signals that are close to black, and you certainly would not want to enhance the detail of the noise component of the signal. To avoid this, a level dependence circuit restricts detail enhancement in near black regions, and 'crispening' is used to inhibit the creation of a detail signal for small transitions in the signal so that noise is not amplified.

Detail correction is normally concerned with enhancing details within an image. But if the presenter is having a bad skin day, or perhaps feels that he or she has a problem with excessive physical maturity, then detail enhancement is not going to be a popular option. The answer, amazingly enough, is to identify areas of skin tone within the image and only apply detail enhancement to the rest. The colour range where detail will not be enhanced is adjustable for phase, width and saturation as shown in Figure 4.5. Note that this is concerned with hue and saturation, which are factors independent of the lightness or darkness of a person's skin.

Although a video camera, by its nature, works with video in its component form; separate or easily separable red, green and blue signals, it has to be recognized that whatever is eventually going to happen to the signal it is virtually certain to be converted into composite NTSC form. It makes sense therefore to apply certain pre-corrections to the signal so that problems due to the NTSC system are minimized, and it is better to do this while the signal is in as pristine a condition as possible, which of course is in the camera. Cross-colour is an effect seen particularly on diagonal stripe patterns and is caused by interference between the luminance (monochrome) signal and the chroma (colour) subcarrier. The result is a multi-coloured pattern that did not exist in the original image. A comb filter is provided to reduce the level of specific bands of

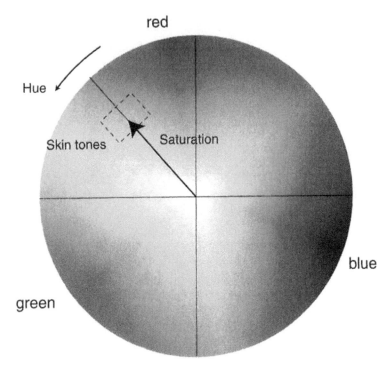

Figure 4.5 Identification of skin tones.

frequencies which can be adjusted according to the degree of cross-colour interference. Since the digital filter has very precise characteristics, interference of this type can be almost eliminated without affecting the overall image resolution.

The DSP capabilities of a high end camera are valuable in correcting the output of the CCDs, which naturally enough can never have totally accurate colour reproduction, or colorimetry, in themselves. It is necessary also to consider that all the way through to the phosphors on the TV screen there are compromises in the things that are technically possible to achieve, and the gap between those technical possibilities and the colour response of the human eye (even then, that's just the average human eye). SMPTE has specified what the ideal colorimetry should be of the red, green and blue channels of a camera (Figure 4.6), and the spectral characteristics actually include areas where a negative response is required. For instance, the green CCD should have a response that is less than zero to blue light. This is of course rather tricky to achieve in the analogue domain, but to a high end camera's digital matrix it is child's play. The matrix is also employed in creating alternative colorimetries, for special effects perhaps, or to match the outputs of multiple cameras.

Although television receivers have a contrast control to adapt to users' requirements, some preferring a bright contrasty image, others more

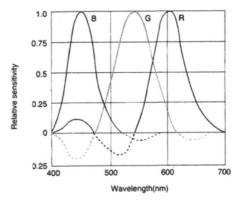

Figure 4.6 CCD spectral characteristics.

natural and lifelike colours, contrast has to be carefully controlled at the source. In the studio it is possible to control contrast by illuminating the scene evenly and not allowing excessively dark shadows to form. On location, and particularly for news, it is a different matter where the contrast of the scene can vastly exceed the capabilities of the television system. Although the camera's CCDs can capture a fairly wide range of brightness values, the end result on your TV screen, without compensation, would be lots of bright areas, lots of dark areas, and not very much in between. To compensate for this the contrast is reduced in the mid-tones and highlights while increasing it in the shadows. The value of gamma ranges between a theoretical 0 (no contrast at all) and 1, 1 being the way the CCD sees the scene. A typical practical value would be 0.45 which shows that the contrast variation in the mid-tones is reduced to less than half. In addition to gamma there is also a knee control which sets the degree to which very bright highlights will be compressed so that they can be accommodated without burning out.

Even when the contrast of the image is set correctly, there may be problems with shading, or uneven brightness, across the image. Shading in dark areas is a product of the thermal characteristics of the electronic circuitry, while shading in light areas is due to uneven sensitivity of the optical system, including the lens, prism and CCD imager. Fortunately, shading in a CCD camera is already less than in a tube camera and the DSP can analyse the image to determine the extent to which shading is taking place, and then automatically generate a compensation signal for each of over twenty-five thousand zones.

Master Set-up Unit

The DSP capabilities of a modern camera obviously provide great flexibility and the images are comparable to the best you will ever see on

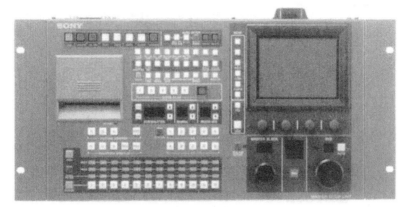

Figure 4.7 Sony MSU-700 Master Set-up Unit.

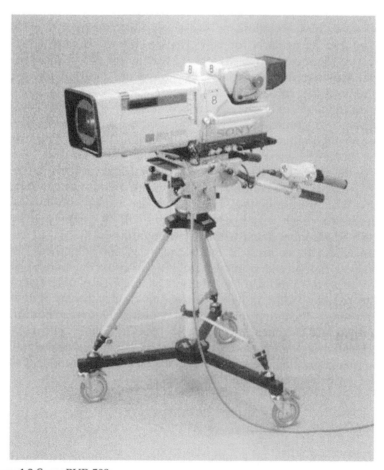

Figure 4.8 Sony BVP-500.

video. But that is just for one camera. What happens when several cameras are used on a multi-camera shoot, as is common for sitcoms and soaps? As I said earlier, although the human eye can adapt easily to different lighting conditions, the camera sees things as they are, particularly the colour temperature of the lighting. When the image is converted into a signal then the eye is much less tolerant of deviations from what it considers the norm and is particularly intolerant of changes in colour balance, as could easily happen when the vision mixer switches from one camera to another. It would be a thankless task for an engineer to have to visit each camera in turn to tweak its adjustable parameters, then compare the outputs and find they were still not quite matched. Going backwards and forwards could take all day, and then what would happen if a camera drifted out of adjustment during the shoot? Bringing

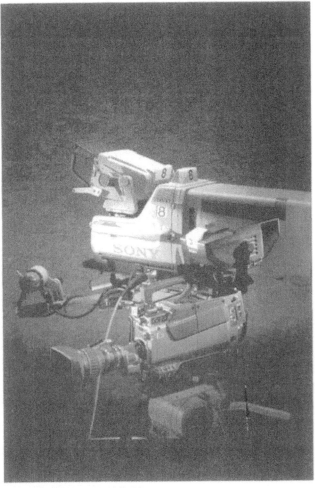

Figure 4.9 Sony BVP-500 and BVP-550.

proceedings to a halt at this stage is going to be expensive. Far better therefore to invest in an important accessory, the Master Set-up Unit. The Master Set-up Unit (MSU) would be operated, not necessarily continuously, by a member of the engineering staff and it governs functions which the camera operator either would not have time to think about, or concerning the technical quality of the picture which he or she is not in a good enough position to judge. The third function, as I said, is to match the quality of the pictures from several cameras. It would be interesting to examine a few of the facilities provided. Remote control over gain can be selected here. All serious video cameras have variable gain controls, which are sometimes automatic, to make up for low light levels. When the lens aperture can be opened no further, the video signal can be boosted electronically which, although it also increases the video noise level, may make the picture more satisfactory subjectively. Sometimes a particular lens aperture may be chosen for its artistic effect, different apertures allowing different amounts of depth of field. By selecting either gain, or by switching in optical neutral density filters, the desired result can be obtained. (Even though these are remote controllable, the camera operator will still like to know about any changes that affect the depth of field. In fact, a message appears in the viewfinder when either the filter or the gain is changed remotely.)

Also on the MSU are three controls for the red, green and blue components of the picture, so that the overall colour can be balanced. Among lots of other interesting knobs and dials are controls for picture

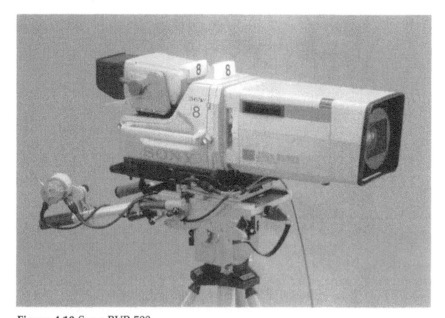

Figure 4.10 Sony BVP-500.

Figure 4.11 Sony BVP-550.

contrast, including the gamma. All the MSU's settings can be stored for later recall, so that complex adjustments may be made for a particular show, and then when that show comes round again the settings can be recalled instantly. There are smaller versions of the MSU, called simply Remote Control Panels, which are more likely to be under continuous supervision. Whereas one MSU can control several cameras, there would be several RCPs banked side by side, one for each camera. The functions are reduced, but even so, the camera operator does not have to go into the studio without someone behind the scenes to take care of the fine technical details.

I have heard it said that the way most camera operators start out on their career is by watching TV as a child and thinking, 'I'd like to do that.' Train companies and fire departments (sound studios too) had better step up their recruitment campaigns!

CHAPTER 5

ENG and Betacam SP

Do you remember what TV news programmes were like in the mid-1980s and earlier? If your recollection is that they were pretty similar to what we have today, then perhaps the slow, steady march of technology has fooled your memory cells ever so slightly. Think hard and you'll recall that the newscasters spent a lot more time on camera than they do now. These days it's hard to catch a glimpse of the latest hairstyles and ties between film reports from the far-flung corners of the globe. In fact, 'film' is the wrong word to use because it was in the old days of TV news – not really so long ago – that 16 mm film was the news gatherers' medium, of necessity rather than of choice. Film cameras have the advantage of being highly portable, but film needs to be processed to be viewable at all. It then has to be edited by cutting the film before being converted via a telecine machine into a transmittable video image. The result was that news was delivered to us, in the main, by the medium of the spoken word, and we had to believe what we were being told.

During the course of the 1980s, we probably did not notice that we were gradually receiving more and more of our news information in the form of pictures from wherever the news was happening, sometimes live, sometimes very soon after the event. Occasionally we were treated to a particularly important item with amateurish camerawork and poor picture quality – but we could still see for ourselves rather than having to take someone else's word for it. The driving force behind this improvement in access to information about what was happening in the world was tersely known as ENG, or more formally as Electronic News Gathering. In essence, this meant the replacement of film cameras by portable video equipment which could offer a greater diversity of news with a much better speed of delivery. Acceptance of this new technology took a little time, as in the case of the newspapers, but now news acquisition on video is the rule, film hardly ever being seen.

Backtracking a little, we ought to examine the role of film in a little more detail in order to see the benefits provided by ENG. Simple 16 mm film cameras were relatively inexpensive and a lot of important events have been captured for posterity in this way. There is, for instance, a good deal of footage of the Hungarian uprising of the 1950s, which is comparable in every way to more recent events in Tiananmen Square, Beijing. Film cameras of the period could be very simple and almost pocketable in size, particularly those which used spring-driven motors and were fully self-contained, requiring only the film stock and enough light to shoot in. But they could only capture pictures. To record synchronized sound to enable a reporter to do the now-familiar talk-to-camera (which provides the important element of on-the-spot analysis of what's happening), the complexity level increases enormously, and the portability of the equipment decreases in inverse proportion. Sound for film can be recorded in the traditional way on a Nagra portable tape recorder which puts a sync pulse on the tape corresponding to the motion of the sprocket holes of the film running through the camera. Back at base, when the film has been processed, the sound is transferred to magnetic film with the same dimensions as the picture film and the two are cut and rejoined in corresponding places. This, as you might gather, is an equipment, time, and labour-intensive procedure, and was always ready for a better way of working to come along.

When video equipment was first introduced to news teams, two time-consuming stages of the process vanished immediately. Videotape does not need to be processed and only needs to be rewound before it can be viewed. Sound can be recorded directly onto the soundtracks of the videotape and is, therefore, automatically synchronized to the picture. Sound can, of course, be recorded and handled separately to some advantage, but speed is of the essence in ENG. Editing can begin as soon as the videotape arrives in the editing suite, and the tape can be made ready for broadcast very quickly.

Betacam

In the early days of portable video equipment, a camera would be linked by cable to a smallish U-Matic video recorder slung uncomfortably over the shoulder, but I shall bypass this transitional phase to get onto the main item of interest, the camcorder. As everyone knows, a camcorder is a video camera with an integral video recorder. The domestic models are great for shooting the kids, and the professional versions are great for shooting news (and even the domestic models find their role when news teams would prefer to be taken for tourists).

Sony's great contribution to ENG was Betacam. Betacam is really a two-part concept, the first being the idea of having a one-piece camera and recorder (often presented as two units which dock together), the second being the on-tape format of the video itself. Betacam is based on the domestic Betamax cassette tape (remember that?), but the video is recorded onto the tape in a completely different way – so don't get the idea that Betacam is an uprated version of the domestic system; it isn't, it just uses the same size of tape. The original Betacam (non-SP) could, in fact, use a standard Betamax cassette, although more highly graded tapes were preferred, so a camera operator could stand a reasonable chance of getting emergency supplies at a Timbuktu hi-fi and video emporium – or they could before the unjustified demise of Betamax.

Where a Betamax cassette could record epic-length films off-air, the maximum running time of a standard Betacam tape is a little over 36 minutes. This is due to a much higher tape speed – 10.15 cm/s. The tape width is $\frac{1}{2}$", as opposed to the $\frac{3}{4}$" and 1" of the current U-Matic and C-Format, respectively. (I should mention, at this point, that Betacam was not the only camcorder that was developed in the early years; there were others, such as the RCA Hawkeye, which used the Chromatrack system of recording onto VHS cassettes.) Betacam SP is a development of the original Betacam format which allows for pictures of full broadcast quality.

The principal difference between Betacam and other analogue formats is that Betacam records component rather than composite video. A video signal consists of a luminance signal, which represents the brightness of the image, and two chrominance signals which describe the colour. These are encoded for transmission into a single NTSC which is decoded at the receiver back into the individual components. The CCD detectors of a video camera generate a component signal in which each colour – red, green and blue – is at full bandwidth and has the maximum amount of detail. This is the ideal type of signal for video processing and, in an ideal world, the picture should remain in component form all the way through the editing and mixing phases right up until the last moment before transmission. Composite video involves compromise and is less than ideal for effects. Other analogue formats write the composite video signal directly, via frequency modulation, onto the tape, or strip out the colour and modulate it onto a lower frequency carrier before recording. Betacam allocates separate areas on the tape for component luminance and chrominance recording and therefore achieves a significant advantage over composite formats.

Compressed time division multiplexed system

The video tracks are recorded diagonally across the Betacam tape, while the two audio tracks, together with the control and timecode tracks, are recorded longitudinally. The luminance (Y) and chrominance (C) tracks are recorded alternately. As explained in Chapter 2, the original red, green and blue components of the picture are converted into a luminance signal (Y) and two chrominance signals from which the three colours can be reconstituted in the receiver. Each track of luminance contains one line's worth of luminance information, and therefore each alternate track of chrominance has to contain one line's worth of each of two chrominance signals. How is this done?

The colour signal from a video camera is inherently in component form; a professional camera will have three CCD detectors, one for the red component of the scene, another for the green, and the third for the blue. This results in an electronic image in RGB form. To record this onto tape would be wasteful, since a full bandwidth signal, 5.5 MHz, would have to be recorded for each of the three colours. Our eyes, however, are much more attuned to fine detail in the brightness (luminance) part of the image than the colour so, if the RGB signal is converted to a luminance and two chrominance signals from which the three colours can eventually be fully reconstituted, then the luminance signal can be recorded at 5.5 MHz, giving a fully detailed picture, while the chrominance signals are painted on with a thick brush at around 1 MHz. Even though the bandwidth of the chrominance signals is reduced, this is still a component signal because, in Betacam, the chrominance signals are reduced to half the bandwidth of the luminance, and then time-compressed so that they will both fit into a single track. This actually gives the chrominance signals extra bandwidth above what is needed for transmission, so the detail contained in the recording is, for most purposes, as good as the signal that came out of the camera.

Since the two chrominance signals have been time-compressed to half their original length, you may well be wondering how they are unsqueezed so that a proper picture can be formed. The answer is that every Betacam player needs to have a form of timebase corrector to assemble the picture with precision registration of all its elements. A timebase corrector stores the information retrieved from the tape, with all its timing irregularities, and buffers it so it can be sent out with the timings of lines, fields and frames exactly according to the specification of a standard video signal. Early Betacam units used CCDs (charge-coupled devices) to achieve this in the analogue domain. Now, of course, it is done digitally.

Audio in Betacam

The Betacam format is, arguably, the first video format to take audio seriously. This is necessary because the whole point of the system is speed of operation from acquisition to transmission, therefore the audio has to be recorded and edited on the videotape itself, rather than being handled separately on synchronized tape or mag film. It makes a lot of sense, therefore, for the audio to be as high a quality as possible. The original Betacam format (pre-SP) used two longitudinal tracks on the upper edge of the tape for audio (there was a separate timecode track on the opposite edge). The tape speed is fairly reasonable for audio, at 10.15 cm/s, but the track width is a measly 0.6 mm which does not bode well for good noise performance – and add to that the fact that the tape isn't optimized for audio anyway. Who could possibly save the situation? Why, Ray Dolby of course, with one of his famous noise reduction systems.

It may come as a surprise that the standard noise reduction system for Betacam audio is Dolby C, a domestic system, rather than Dolby A (SR wasn't around at the time of Betacam's introduction), but it all comes down to portability. A Dolby A processor card – even a miniaturized version – is still a bulky item, far too big for a camcorder with masses of video circuitry to fit in too. Dolby C, on the other hand, had been integrated into chips for the massive consumer market and was available at just the right physical size. Also it offered around twice the noise reduction capability (though, perhaps, not at the same level of quality and not at low frequencies), and it fitted the bill pretty well. Looking at the specification of a modern Betacam unit, you will find an audio performance with frequency response from 50 Hz to 15 kHz (+1/–2 dB) and a signal-to-noise ratio of more than 68 dB on metal particle tape, which isn't bad at all considering.

Betacam SP improved the performance of the luminance and chrominance signals and also added a new type of audio recording – frequency modulation. Two channels of audio are modulated onto FM carriers and recorded on the same tracks as the chrominance information. This causes very little conflict. The audio carriers and their sidebands fit comfortably underneath the spectrum filled by the chrominance signal. The performance of these FM tracks is rather better than the longitudinal tracks with a frequency response of 20 Hz to 20 kHz (+0.5/–2.0 dB) and a signal-to-noise ratio of more than 70 dB. Wow and flutter is specified at an excellent 0.01 per cent. The main drawback of the FM tracks is that they cannot be edited independently of the video information, since they are permanently bonded to the chrominance signal, but they are very well suited for acquisition purposes since, the higher the quality you start off with, the better the end result will be, even if subsequent stages do not have the same level of performance.

Not content with adding 'hi-fi' audio tracks to Betacam, Sony has gone a step further still, and has found space to fit in a pair of digital audio tracks on the latest machines. Amazing but true, considering that digital audio on videotape would have been nothing but a dream at the time of Betacam's introduction. Unfortunately, one of the old longitudinal analogue audio tracks has had to be sacrificed but, when you think of the potential benefits, who cares? The digital audio information is slotted, at the end of the video scan. The sampling frequency is selectable for 48 kHz and 44.1 kHz and the number of bits is 16, making the audio quality equivalent to dedicated digital audio recorders. Since the audio is recorded in a discontinuous fashion, it is evident that it, like the chrominance information, will have to be time-compressed and then re-expanded. But will this cause any strain on the storage medium because of the very high frequencies involved? As it turns out, the digital audio information is recorded onto the tape at a frequency of 8.5 MHz which is high, but no higher than the frequency of a peak white signal in the luminance video channel. The only compromise that has had to be made is that the form of error correction that is most suitable for digital audio does not lend itself to the video editing environment where cuts are made on the frame boundaries. Nevertheless, the Block Cross Interleaved Reed-Solomon code employed can deal with a burst error equivalent to eight lines of picture, which should cover most cases.

The future

Betacam SP is, and will remain, a very significant format for several years to come. If Sony were to discontinue Betacam SP production today, then the vast amount of equipment currently in use would be lovingly maintained well into the future. In fact, I wonder whether the market will ever allow Betacam to fall into disuse as it is almost as universal a standard in video as 35 mm is in film. Obviously, digital video is the way of the future, once we have got past the current confusion of formats and settled on a dominant standard (but when will that happen?). Right now, however, Betacam SP can be considered to be future-friendly. The most important feature of Betacam SP is that the signal is in component format and it can be copied onto a digital medium perfectly easily. In fact, the

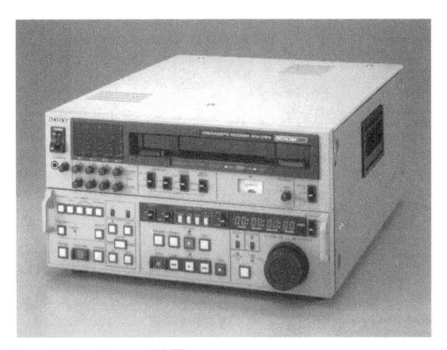

Figure 5.1 Sony Betacam SP VTR.

latest incarnations of Betacam – Digital Betacam and Betacam S – are both backwards compatible with Betacam SP, and you can buy a Digital Betacam or SX machine that will replay an SP tape without any problem. This means that any organization with an investment in SP camcorders can continue to use them until they eventually wear out.

It has to be said that Betacam SP is not without faults. When a well cared for unit is used for drama production, the results are very nearly as good as anything the digital formats have to offer. In news acquisition, Betacam SP camcorders get a hammering and, particularly when tapes are recycled, the result is a glitching of the picture, sometimes mild, sometimes severe. Betacam SP has served us well, however, and in a few years time when it is accorded 'classic' status, if such a thing exists in video where people are rather less sentimental, it will surely deserve it.

CHAPTER 6

Digital video

There was much scepticism when digital audio recorders were first introduced. The sound was harsh ... the sound was cold ... editing was difficult ... the equipment was expensive – a whole list of complaints from those with a heavy investment in analogue recorders, and others. Where are the sceptics now? They are still there but are very few in number and most have come to agree that digital audio offers so many advantages for the storage and manipulation of sound that its few disadvantages can be almost completely disregarded. Even if you prefer the sound of analogue tape, it is almost certain that the last resting place of your signal will be on a digital medium. Strangely enough, there never was any similar complaint about the introduction of digital video recorders. Perhaps the concept was easier to accept after the introduction of digital audio, or perhaps the shortcomings of analogue video recorders are such that the professional user has longed for a superior technology to take over. Digital video processing has been with us for some time in the timebase correctors necessary for successful C-Format analogue recording, but actually storing digital video information on tape proved rather more difficult. Seemingly insurmountable problems, to engineers, are the source of a great deal of job satisfaction, and the problems posed in achieving a data rate of more than 200 Megabits per second must have been particularly mouth-watering (compared with the upper frequency limit of analogue video, a mere 5.5 MHz, which took many years of research to achieve reliably). But the advantages of digital video in the form of rock-steady pictures and the complete absence of generation loss in editing and distribution are considerable and in the twelve years since its introduction it is now close to ousting analogue machinery totally for both professional and amateur user.

D1

D1 seems an appropriate name for the format used by the first commercially available digital video recorder. The D1 format was finalized in 1986 and the first machine to be introduced was the Sony

DVR-1000. Many aspects of the format revolve around the need to cater for both 525 and 625 line working as most of the world outside North America uses a 625 line format. Since D1 is a component recording system it is not necessary to have different models for 525 line NTSC and 625 line PAL – one recorder can handle both. Even though there is no composite coding to worry about, the difference in the number of picture lines in the two systems needs to be accommodated. As in digital audio, the analogue video signal from the camera is sampled and quantized into discrete steps so that the information can be stored as a sequence of numbers. In digital audio, the sampling frequency is usually 44.1 or 48 kHz – pretty puny compared with the sampling rate of 13.5 MHz used for the luminance (brightness) signal in D1! The chrominance signals do not require such a high bandwidth. In D1, the chrominance (colour) signals are sampled at 6.75 MHz. Put simply, this means that the recorded luminance signal is capable of twice the resolution of the chrominance signals, which fits in with the way the human eye works and still adds a safety margin of colour detail.

It may come as a surprise, seeing that everything about video seems to involve higher numbers, that the signal is quantized to only 8 bit resolution, compared with 16 bit for DAT and CD, and now 20 bit or 24 bit for high end audio. Sixteen bit resolution means that 65 536 separate levels are measured and encoded; 8 bit resolution results in only 256 different levels, which may seem to be a shortcoming but it does appear that, for TV, analogue video signals and 8 bit quantized digital signals are indistinguishable to the eye. In fact, in the heyday of analogue video recorders many timebase correctors were 8 bit digital so we have been happily watching digitized video signals on our televisions for years without any complaints. Although 256 levels are available, in the D1 luminance signal 220 steps cover the range from pure black to pure white, the remainder being left available to accommodate peaks which may occur from camera signals or D/A conversion.

It is interesting to see how some of the figures involved relate to the NTSC or PAL systems. The 13.5 MHz and 6.75 MHz sampling rates were chosen so that there will be a whole number of samples per line in each system. The way the tracks are recorded on the tape also involves some interesting numeration. To record a complete 525 line frame at a frame rate of 29.97 Hz, 20 diagonal tracks are recorded on the tape. To record a 625 line frame at a frame rate of 25 Hz, 24 tracks are recorded. $29.97 \times 20 = 599.4$, $25 \times 24 = 600$ – figures which are so close that in either format 600 tracks are recorded every second. Very clever.

One question which may have arisen by now is why should the first digital format have been component rather than composite? After all, it seems that it must be more complicated to record component signals, and a new machine in this format would not be a direct replacement for the one inch C-Format recorders in the predominant composite video environment. The answer to this is that the feeling at the time was that

C-Format was functioning very satisfactorily and digital technology would be a giant leap into the unknown – it had to offer something different, and then prove itself, before it could be accepted elsewhere in the broadcast environment.

D1 error protection

Several techniques are employed to reduce the possibility of a drop-out on tape affecting the reproduction. The first is a very simple binary number trick known as video mapping. If the signal were encoded into simple 8 bit binary (for example, 01111111 = 127, 11111111 = 255, 10000000 = 128, 00000000 = zero) then if an error occurred in the first digit, the value would be out by 128, as you can see from the examples. This would very probably be visible on screen. To avoid this a table has been constructed such that consecutive binary numbers are mapped onto a non-consecutive sequence of numbers arranged so that any errors in recording and reproduction will have minimal effect. On replay, the mapping process is reversed of course. Video mapping is a separate thing from error correction, and indeed if error correction always worked perfectly then mapping would be irrelevant.

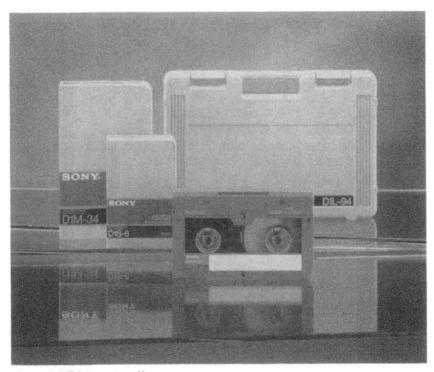

Figure 6.1 D1 tape cassettes.

Figure 6.2 Sony DVR-20 D2 recorder.

Figure 6.3 Sony DVR-2000 D1 recorder.

It is fairly well known that the data in digital audio is not written sequentially onto the tape but interleaved so that if there is a drop-out it will not kill one section of audio totally, but cause smaller errors – with a wider distribution – that can be easily corrected. The obvious way to spread the burden of error is to create a simple structure of interleaving, which is fine for audio, but not so good for video since errors might be seen as regular patterns on the screen which will be noticeable to the eye. In D1, a shuffling process is used which disperses the data more irregularly, although always within limited confines since there is also the requirement for the recovery of some sort of picture in shuttle mode. Further error correction is incorporated in the channel coding itself and although the products of the top manufacturers are robustly designed, they have to work within a very tough environment and effective maintenance is still vital.

Well after its launch, D1 is still going strong in post-production. It isn't the only component digital format any more but much of the product you see on your screens, particularly where effects are concerned, will have been through a D1 recorder.

D2

Whereas the first D1 machine was a product of Sony's research and development, D2 began as an Ampex initiative to replace their automated broadcasting system using cartridges of 2" tape in the Quadruplex format. It is important to realize straight away that D2 did not replace D1, nor does it offer higher technical quality. The main advantage of D2 was that it could fulfil the same function as a C-Format recorder by replacing it within existing analogue composite installations with a machine with a better picture and reduced running costs. Since D2 is a composite format there are the two mutually incompatible standards of NTSC and PAL to be considered, so D2 machines are available in two versions. The technical quality could never have been as good as D1 because the imperfections of the composite video encoding process are faithfully recorded and reproduced, but advantages over analogue composite recording, such as C-Format, include the absence of moiré patterns due to FM recording and a timebase error of zero. It is an interesting point that although timebase correctors in analogue formats such as Betacam SP can do a good job of pushing around the lines and fields into their correct positions in time, they can do nothing for chrominance phase integrity during individual lines because the phase reference, the colour burst, occurs only once before the line commences. Degradation of the chrominance information in the picture over several generations of copying is a particular problem with analogue composite video.

Although D2 is a digital format and in an all-digital studio generation loss is a thing of the past, the key point about D2 is that it had to function well in a predominantly analogue environment. One result of this is that the sampling frequency is much higher than is strictly necessary according to sampling theory. In the PAL D2 system this works out at four times the colour subcarrier frequency, or 17.7 MHz (once again, compare this with the 44.1 or 48 kHz of digital audio). The advantage of this is that when the signal is repeatedly encoded and decoded from analogue to digital and back again, as it inevitably will be in an analogue-oriented edit suite, the filters do not have to be as steep as they would otherwise have been and pass band ripple (an irregularity in frequency response) is minimized.

On tape, D2 uses the azimuth recording technique where two heads are set at slightly different angles and write adjacent tracks on the tape with no guard band as found in D1. Each head responds only to tracks recorded in its own azimuth angle. Like D1, D2 is a segmented format where the information for each field is spread over more than one track on the tape. In analogue video this can cause problems of picture banding when the heads are not absolutely precisely aligned, but with digital recording there is no such disadvantage to segmentation. A complete field takes up three segments, each of which consists of a pair of tracks, one in each azimuth angle. In shuttle mode, when the tape is moving faster than normal and some fields need to be omitted, the heads effectively have to jump from one position on the tape to another. When the shuttle speed is slightly higher than normal, the heads can jump during the audio blocks at the end of the track and settle during the audio blocks at the beginning of the next track. As the shuttle speed increases, more jumping and settling time is needed and the heads can only pick up the centre part of the video track successfully. Fortunately, due to the design of the format, the outer sections of the video track contain redundant information for use by the error correction system, so a complete picture can be obtained from only the centre part of the track.

D3 and D5

As you might expect, improvements on the standards of the original digital video standards were not long in coming, and Panasonic were the company to do it. In a previous generation, Ampex and RCA were the big guys of broadcast video, then it was Ampex and Sony, then Sony and Ampex (note the change of emphasis) and now Sony and Panasonic. The initial thrust for D3 came from the major Japanese broadcaster NHK where they could see an end to the life cycle of their 1" equipment and wanted a modern replacement. As I said earlier, D2 was a very suitable replacement for a 1" machine, but only in the studio. Both D1 and D2 formats use $\frac{3}{4}$" tape, which is too wide to incorporate into a camcorder format. Panasonic designed the D3 composite format to use $\frac{1}{2}$" tape. A quarter of an inch may not sound like that much of a difference but apparently it all adds up. D3 therefore is as much a format of acquisition as it is post-production, and a recording can be made in a D3 camcorder, sent to the studio and slotted directly into a studio D3 machine for editing. The other major difference between D2 and D3 is that where D2 works to 8 bit resolution, D3 is 10 bit, allowing 1024 levels of encoding. The advantage of 10 bit resolution is that although 8 bits are enough to display a very good image on a typical TV screen, they are only just enough and there is no margin either way. The blackest black on the tape is only as dark as you see it on your screen. Likewise, there is no signal brighter than the brightest signal your TV can display. This implies firstly that there is no margin for error when shooting on an 8 bit format, and also that there is no leeway for adjustment of the shadows and highlights of the image in post-production. This is where 'old fashioned' film scores, where there is still an immense advantage even over the latest digital video systems.

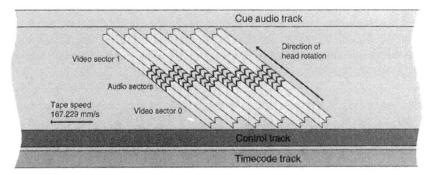

Figure 6.4 Track layout in D5.

Panasonic were thinking ahead when they designed D3, and already had their sights set upon a 10 bit component format, which eventually saw the light of day as D5. If you are wondering what happened to D4, the word for four in eastern Asian cultures sounds very much like the word for death, and is therefore considered unlucky. D5 can be considered an improvement on D1 in that it records component video to 10 bit resolution rather than D1's 8 bit. Unfortunately for Panasonic, D1 had already become something of a standard and the acceptance of D5 was slower than it might have been. It has been said that if Panasonic had come in earlier with their systems they might have swept the board and become the major supplier of broadcast video recorders, but Sony were already established and Panasonic have had to work hard to find niches to sell into, despite their technical excellence. Interestingly, D3 and D5 are compatible to a certain extent as it is possible to buy a D5 machine that will play back D3 tapes. This idea of backwards compatibility will continue to be important as new formats are developed. Which leads to the inevitable question . . .

Why so many formats?

Digital video doesn't stop at D5, as you know. The analogue formats were comparatively long-lived and in broadcast terms there were only three major formats covering four decades (Quadruplex, C-Format and Betacam SP). But the difference between analogue and digital video is that analogue is capable of incremental improvements; the image and sound quality of a digital format is carved into a tablet of stone when the design committee have finished their discussions and before the first piece of equipment is sold. It is no use a tape manufacturer coming along saying that you can get 3 dB more output or whatever. This might improve the error rate off-tape but it won't improve the picture quality under normal operating conditions. If you consider analogue audio, the performance of $\frac{1}{4}$" tape machines improved vastly over their forty years of currency and – if you discount the use of noise reduction systems – it is still the same format. There will never be a digital format, audio or video, that lasts forty years. We, whether as manufacturers, producers or consumers, will always want that extra bit of quality that new technology has made possible, and that will mean a new format. This is the new way of things so we have to learn to love it.

Composite and component

In the video camera, three CCD sensors produce signals for each of the red, green and blue components of the colour image, being known collectively as an RGB signal. To transmit or record each of these as a full bandwidth signal would be wasteful since the human eye is less sensitive to fine detail in colour than it is to detail in brightness information. The RGB signal is therefore converted to a high resolution luminance signal (Y) and two lower resolution chrominance signals, U and V (I and Q in NTSC) which together make up a component video YUV (YIQ) signal. In this component form the signals are kept separate and can easily be reconstituted into RGB if necessary. However, they need three coaxial cables, or equivalent, to get them from one place to another. It isn't convenient to transmit YUV or YIQ as it stands over the air so the signals are converted to composite form according to the PAL or NTSC method. Composite video is convenient to handle, and can be carried by one coaxial cable, but it has disadvantages such as patterning caused by crosstalk between the chrominance and luminance signals, and in analogue composite recorders another patterning effect known as moiré caused by the use of the frequency modulation recording technique. Component signals are also more appropriate for effects. The D1 and D5 formats record component signals directly and provide a very high quality picture without the limitations of PAL or NTSC encoding.

DV and DVCPRO

DV, short for Digital Video, is one (just one) of the more recent digital video formats. But it is not a professional format: DV is a domestic camcorder format and, yes, you can now throw away your SVHS-C or Hi-8 home camcorder because it is now hopelessly out of date. The biggest news as far as the consumer is concerned is apparently that since DV is supported by fifty-five major manufacturers, including JVC, Panasonic and Sony, then there should be no format war in this round of technology. Personally, I wouldn't bet on it, but let's wait and see. 'But hang on a minute,' you say, 'I thought that *A Sound Person's Guide to Video* was about professional video?' And so it is, but to bring a new line of products to market these days, a big company needs all the support the massive domestic market can offer, and any similarities between domestic and pro technologies are purely intentional. Although DV is a domestic format, you can be sure that it will be used for professional purposes, and to make sure of a healthy share of that market too, Panasonic have produced a beefed-up version of the DV format and called it DVCPRO (Sony did a similar thing with DVCAM). DVCPRO equipment uses the same basic format as DV, but measures have been taken to improve the robustness, up to broadcasters' stringent – almost military – requirements. But let us look at DV first, and absorb the basics of this new technology.

Small is beautiful

If you think that DAT cassettes are small, wait till you see a DV cassette: $66 \times 48 \times 12$ mm are the dimensions. In fact the extra thickness compared with DAT brings the volume of the two cassettes up to around about the same, but when you consider that a DV cassette of this size (actually the smaller version is called a 'MiniDV' cassette) can store an hour of video, plus two channels of 16 bit 48 kHz audio, then the words 'mind' and 'boggling' seem to be perfectly appropriate. To put it another way, if all the data on a high density floppy disk were packed as tightly as DV data, the diameter of the disk would be just a little over 6 mm! A bigger cassette is also available that allows 270 minutes of video with an extra two channels of audio. Even so, the size is only $98 \times 64 \times 15$ mm, which is a fraction of the size of the larger cassettes of other digital formats. (I should point out that in professional video it is common for a format to offer multiple cassette sizes, typically small, medium and large.)

Of course, it would not be reasonable to expect to be able to capture every last video digit over such a long duration into such a small package, so video compression has to be employed. As a starting point, the video data consists of images 720 pixels wide by 480 high. The images are stored in component form (which keeps the colours in a purer form than PAL, NTSC or SECAM encoded composite video) at 8 bit resolution.

Luminance (monochrome brightness information) is sampled at 13.5 MHz. Chrominance (the colour) is sampled at 3.375 MHz which is, as you may have noticed, a quarter of the luminance sampling rate. It is normal in any video format to allocate a lower bandwidth to the chrominance signal to save on storage space and data rate. This matches the requirements of the eye very well, although DV's 4:1:1 ratios between the sampling rates of the luminance and the two chrominance signals is rather more of a compromise than the 4:2:2 of most of the other digital formats, including another popular compressed format, Digital Betacam.

JPEG (Joint Photographic Experts Group) image compression is designed for still images and MPEG (Motion Picture Experts Group) obviously for moving pictures. JPEG has also been adapted, as Motion-JPEG or M-JPEG, for nonlinear video editing systems since it has the advantage of encoding every frame separately, which is obviously very appropriate for editing. MPEG on the other hand only fully encodes a certain number of frames, and in-between frames are described in terms of how they differ from the key frames. It is more difficult to apply MPEG to editing systems, although certainly not impossible. DV uses the same basic Discrete Cosine Transform compression of JPEG and MPEG at a ratio of 5:1, but doesn't go as far as MPEG would in only partially encoding frames. According to the requirements of the image, the encoder would compress video fields (a field consists of either the odd or even number lines of a frame) independently, or it might combine two fields together and compress them together. The data rate is reduced by these means to a manageable 25 Mbits/s. Obviously, after all this compression you would not expect the image quality to be up to the standard of a non-compressed 'transparent' format like Panasonic D5, although some say it is superior to Betacam SP; but as we shall see, DV's and DVCPRO's real advantages lie elsewhere.

Nuts and bolts

I think there will be a label on DV equipment saying, 'No user serviceable parts inside', and this time it will mean it. The 21.7 mm diameter head drum for instance rotates at 9000 rpm which is faster than most hard disk drives. The twin heads lay down twelve diagonal tracks per frame, in a similar fashion to all other helical scan recorders, a mere 10 microns apart, closer than DAT in fact. Video, audio and subcode data are kept separate with 'edit gaps' so that insert editing can be performed on picture and sound independently. Error correction is said to be robust and apparently two whole tracks can be lost per frame because of drop-outs and the entire data for the frame can be reconstructed from what is left.

The tape itself is 6.35 mm – a quarter of an inch – wide and has a metal evaporated coating, similar to that used in some types of Hi-8 cassette.

Sony DV cassettes incorporate a memory IC which can store a table of contents and other information which might be useful at a later data. Currently, Panasonic's cassettes do not have this chip, and although both cassettes are mutually compatible, it is debatable whether this is sowing the seeds of some kind of divergence.

Many DV camcorders incorporate the IEEE 1394 Firewire interface which allows data to be piped directly into a computer, and out again. As well as video data, timecode and edit control data can be delivered in the same way.

DVCPRO

As I mentioned earlier, DVCPRO is Panasonic's development of the DV format which is intended to make it more attractive to broadcasters, although plain DV will probably do pretty well in this market too. The main differences are these:

- Although metal evaporated tape offers a better data packing density than other tape types, there is still some doubt over its robustness. DVCPRO uses metal particle tape, like other professional digital video recording systems.
- An extra analogue audio track is provided to allow audio playback in shuttle mode, or to act as an additional separate low quality track if need be.
- An extra control track allows faster servo lock after mode change, for instance from stop to play. It also allows a shorter pre-roll time in editing.
- Tape speed and track pitch are both increased. The reduction in recording density in going from a 10 micron track pitch to an 18 micron pitch offers greater reliability.
- The subcode area of the tape is used to record LTC and VITC. Both types of timecode can be read at any speed.
- Optional serial digital interface.
- The 'C' in DVCPRO stands for 'cassette'. It was dropped from DV. (In case you were wondering!)

The greater ruggedness of the DVCPRO format will set it apart from DV, but both will certainly do well in the professional market. Remember that DVCPRO VTRs can play DV tapes without modification. The size of the cassette is going to be the key selling point, since all the other advantages revolve around this. In audio, many people think that DAT cassettes are too small, and this point of view can easily be justified. In video, there is a serious point in making equipment as compact as possible, so DV is not just small for the fun of it – a smaller cassette means a smaller camera. The current generation of broadcast camcorders are small compared with the previous generation of separate cameras and recorders, but the loading on the camera operator's shoulder is considerable. Even if it is thought acceptable to load the camera operator with as much as they are able to bear, then mobility is restricted with a consequent impact on acquisition opportunities. The cost of air freight, too, can be reduced if equipment is lighter and more compact, considering that news teams take practically a complete studio with them these days, so that they can edit their stories on the spot and send them home over a satellite link. Panasonic are very well aware that equipment does not just cost what it costs to buy – a broadcaster needs to take into account the total lifetime costs of purchase, including maintenance, tape supplies, and even distribution and storage costs of the tape. A typical ENG camcorder may cost around £40 000 with lens. The DVCPRO equivalent should halve this. Savings on tape costs should be around 35%.

A compact cassette has a knock-on effect all the way through compact cameras, VTRs, editing equipment and smaller and lighter flight cases. If a camera can be smaller and lighter, and considering that the format originates in the domestic market, then it can be cheaper, so it is a real possibility to give cameras to journalists and stringers. The man or woman on the spot will have a chance to capture the action, even if it is not quite to the level of skill that the camera crew will be able to offer when they arrive – but it may all be over by then. It is interesting to consider that once upon a time a news crew would have consisted of a director, camera operator, presenter and sound recordist. Now one person has to fulfil all of these roles. Do they get paid four times as much, I wonder? You will be aware that it is already fairly common to provide journalists with Hi-8 cameras, which are almost cheap enough to be regarded as disposable (perhaps one day the single-use camcorder will be with us!). But the picture quality is lacking compared with professional quality equipment. Now broadcasters will be able to take advantage of the superior quality of DV and DVCPRO with video journalist camcorders. These are 'palmcorder' type ultra compact units costing around £3000 with three CCDs compared with the normal domestic camcorder's one. News organizations, both print and electronic, are buying in bulk.

Yet another advantage of the small cassette size is the possibility of designing types of equipment that have not been practical before. One example of this is a Laptop Field Editing System which incorporates two DVCPRO VTRs with twin colour LCD screens and edit controller, making it possible to carry the unit onto a plane as hand luggage and edit your story on the flight home! Other equipment in the range includes a quadruple speed transfer player. Nonlinear editing is now fairly widespread, but still suffers from the drawback of having to copy the contents of a tape onto the hard disk before any work can be done. Earlier generations of digital recorders have only been comfortable with the data rate at their normal play speed, and although fast shuttle is possible, image quality is degraded. DVCPRO is a compressed format so the data

Figure 6.5 Panasonic DVCPRO50 range.

rate off tape is lower, which makes it practical to run the tape faster to shift the data at a greater rate. All of this extends the capability and efficiency of the system as a whole.

By my reckoning, the world now has all the video formats it could possibly need, DV and DVCPRO having filled what must surely be the last gap. But of course I know I'm wrong, the creativity of video equipment manufacturers seems to be boundless, and in six months time (whenever you are reading this) there will be another new format. Don't you just love how calm and sedate audio is in comparison?

Digital cinematography

'The look of film' is a phrase that reverberates around the corridors of production companies and broadcasters worldwide. In four short words it defines the cultural gap, or should I say chasm, between chemical images and those captured electronically or digitally, between film and video in fact. Film is associated in our minds with art and spectacle, with people who strive to provide the best in the world's most expensively produced form of entertainment. Video on the other hand is associated with news reports, soap opera and camcorder shots of the kids. It is not surprising that those who work with video want to emulate film so that a little of the essence of its cultural associations can rub off. But if film is so great, why choose to shoot on video in the first place? The most significant reason is that when you shoot on video, the bottom line cost is

DVCPRO50

Panasonic's DVCPRO50 is designed to remove the perceived problem of conventional DVCPRO's 4:1:1 digital data format. '4:1:1' simply means that the two chrominance signals from which red, green and blue information is decoded are sampled at a quarter the rate of the monochrome luminance signal. '4:2:2' means that they are sampled at half the luminance rate. 4:2:2 is considered fine and beyond what can be transmitted or appreciated while the ultimate of 4:4:4 is only of relevance in effects work. The reduced bandwidth chrominance signals and 5:1 data compression of DVCPRO result in a data rate of 25 Mbit/s, compared with more than 200 Mbit/s in the early D1 digital video format. Although video can be compressed much further, to the point where it has a lower data rate than audio of any listenable quality, the degree of compression has been wisely chosen so that artefacts are hardly ever visible, and compression is always 'intraframe', which means that data is never cross-referenced between groups of pictures. Intraframe compression is fairly obviously significant when images are to be edited. For DVCPRO50, the data rate has been doubled to 50 Mbit/s which allows full 4:2:2 encoding and a gentler compression ratio of 3.3:1. This is achieved by running two DVCPRO chipsets in parallel and by increasing the tape speed from 33.8 mm/s to 67.7 mm/s (compared with 18.8 mm/s for DV). Standard DVCPRO tapes can be used, although the recording time is halved from a maximum of 63 minutes down to 31.5 minutes for the medium cassette size, and from 123 minutes down to 61.5 minutes in the large. A longer tape is planned which will allow greater duration.

---

lower. The rental cost of the camera may be as high, but the stock costs can be vastly less – around 5% of the cost of the equivalent quantity of 16 mm film, and that doesn't even take into account the processing of the film, production of rushes, colour grading and the other necessities of chemical imaging. Once upon a time, film was considered to be easier to edit than video, which would result in a better paced end product, but nonlinear editing has eliminated this advantage. But when you have enjoyed these cost savings and your production is ready for transmission, what will you have lost that could have been achieved if you had shot on film?

Probably the most significant loss is that a video recording looks like a video. It is hard to define, but even in a very good quality video image we see things that remind us of the old days of video when defects were very visible. This has much to do with the way a video camera handles gradations of tone and colour, and although a broadcast quality camera may do a very good job technically, the result might not be as pleasing to the eye as film. Another important factor is that a video camera does not have as wide a 'dynamic range' between shadows and highlights as film does. Film can capture a very much wider range than can be displayed on a TV receiver, therefore it allows the possibility to adjust the tonal range after shooting. With video, there is less of a margin for adjustment and care has to be taken to shoot the pictures as they are intended to be viewed. Image quality is not all that is lost, however. If a drama is shot on film, and it turns out to be exceptionally good, there is the possibility that it might be shown in the cinema. A 35 mm print can be made from a 16 mm negative, and though the quality will not be as good as a 35 mm original could have been, few members of the audience will consciously notice the difference. It is possible to transfer video to film, but the quality gap is hard to bridge. Shooting on film also makes sense if you want to sell your programme into territories where a different TV standard is used. Film will convert equally well to NTSC or PAL. Converting NTSC to PAL or vice versa is of course possible, but it means that you end up with the defects of NTSC, the defects of PAL and the defects of the transfer process.

Having said all that, people still want to shoot on video, but they want to achieve results which are artistically as good as film. One possible solution to this is to try and emulate the defects of film, which is by no means a perfect medium. Another option is to attempt to bring video to a state that is as close to perfection as possible, so that film might well be seen as inferior. This is where Digital Cinematography comes in. You might well ask whether this is anything new, or just a marketing buzz phrase. We have had digital video recording equipment available since the second half of the 1980s but I think it is correct to say that it has always been seen as coming from the video culture rather than seeking to emulate film culture, which is a rather different thing. The early digital recorders were bulky and unwieldy devices anyway (even compared with a 65 mm Panavision camera!) and were for the most part studio-bound and unsuitable for location work. Digital Cinematography is all about the camera, and what was needed to make the concept work was a digital camcorder, which we now have in the form of Digital Betacam.

Step up from SP

Digital Betacam is the next logical step from Betacam SP. Betacam SP, to give you a little history, is a development of the original Betacam which was a professionalized version of the old domestic Betamax system. Betacam indeed used $\frac{1}{2}$" Betamax tapes which at the time were fairly widely available and easy for a news team to acquire almost anywhere. Betacam SP took advantage of new tape formulations with a higher magnetic 'packing density' to achieve a picture quality of full broadcast standard. In both Betacam and Betacam SP, the small size of the cassette allowed combined camera recorders, or camcorders, to be manufactured, something that to my knowledge has never been achieved with the $\frac{3}{4}$" or wider tape of other formats. (Panasonic also make camcorders in their $\frac{1}{2}$" digital formats.) Some people will say that although a new Betacam SP camcorder can record broadcast quality images, any wear and tear will soon degrade the quality, and you only have to look at news reports to see how degraded the quality can become. To be fair, some of the glitching that you will see is due to the rough handling the equipment receives, and the fact that tapes are usually recycled. Drama should always be shot on carefully maintained equipment on new tape.

When Betacam SP was new, its competition was the old analogue C-Format, which is now generally only used to play archive material. C-Format was acknowledged as having generally better picture quality than Betacam SP. But C-Format is a 'composite video' format where the signal is encoded into PAL or NTSC. This process, where the colour information is interleaved into gaps in the frequency spectrum of the monochrome signal, isn't perfect and there is a certain amount of interference between the chrominance and luminance signals. In addition, although the composite format is fine for editing and simple transitions such as dissolves, it is not ideal for more complex effects as there is no direct access to the red, green and blue signals without separating them out and then re-encoding. This of course would magnify the deficiencies of the NTSC and PAL processes. The alternative to composite video is component video, where the luminance and chrominance signals are not mixed together but handled and stored in a form where the primary colours can be extracted without loss. A video camera is inherently a component device producing one signal for each of the primary colours. At the other end of the transmission chain a television receiver or monitor is also component since it has physically separate phosphors for red, green and blue. Obviously, since composite video is inferior in terms of image quality, in an ideal world it would be used only when strictly necessary – for transmission. The rest of the signal chain would be component. This in fact was one of the strengths of Betacam SP, that it was, and of course still is, a component format. Component signals from a Betacam SP tape are ideal for effects, and also transfer well from PAL to NTSC and vice versa, since the signal only needs to be encoded once. (Betacam SP has NTSC and PAL options because of the difference in the frame structure, 525 lines at 30 frames per second as opposed to 625 lines at 25 fps.)

Betacam SP was not the only component format around, pre-Digital Betacam. The D1 format, developed by Sony as far back as 1986, is also component and has been a valuable tool for high end work in graphics and effects ever since. Since the D1 format is now looking rather elderly you might expect it to have been overtaken by newer technologies in some areas, and there are a few aspects where it is now open to criticism, while acknowledging the fact that it was an incredible achievement in its day. It is fair to say that in the early 1990s two quite different areas of the video world were looking for a technology update. Betacam SP users wanted something that had all the SP advantages such as portability and universality, but with an image quality not so liable to criticism when it is having an off day. D1 users staggering under the burden of their finance payments wanted something a little more affordable with the recent facilities D1 lacks. SP users will certainly be pleased with the digital quality of Digital Betacam. Digital recording means freedom from SP-type glitching, at least until conditions worsen to the point where the error correction process cannot cope any longer. Encoding is to ten bits rather than D1's eight which allows for around four times as many different levels of brightness to be recorded. Other factors such as 20 bit audio, stunt playback modes, faster shuttle speed, reduced stock and storage costs make Digital Betacam a very attractive alternative.

Compression

D1 die-hards will point to one potential drawback in Digital Betacam, the fact that it uses compression to store the vast quantities of data in the digital video stream on a comparatively small area of tape. Video compression is something to which we are becoming increasingly accustomed. All the nonlinear editing systems use compression to pack a longer duration on the hard disk, and also to reduce the speed at which the data has to be retrieved from the disk. Top of the range nonlinear editors now offer a maximum image quality afforded by roughly 2:1 compression. One assumes that uncompressed storage must be the goal, but it is really very difficult to find any fault with 2:1 compressed images, particularly moving images. This applies to Digital Betacam too. It would be interesting to compare Digital Betacam and D1 side by side. I would expect both to look very similar, except that the Digital Betacam image would have a wider dynamic range which would allow for a greater degree of adjustment of brightness or contrast in post-production. I very much doubt if you would see the effects of compression.

This is not to say that D1 is superseded. For critical graphics or effects work where several, or many, layers of effects are employed, then each time the signal is uncompressed and then recompressed there will be some degradation, which will obviously build up generation after generation. D1 is a so-called 'transparent' format and any degradation would be solely due to the effects devices, provided the signal stayed in digital form and was not subject to error concealment. Also, outside of graphics and effects work, in an extreme case where part of the image was composed of a very random signal but the rest was normal, for example a scene including a TV set displaying a noisy image, the random signal might push the compression system over the edge and the whole of the image may become blocky, becoming most noticeable in the normal part of the scene. It might be interesting to telecine a couple of choice scenes from the film Poltergeist onto Digital Betacam and see what results! Poor handling of noisy signals is common to all image compression systems, not just Digital Betacam.

Yet another format?

How many times will we ask this question in our careers? Or will there one day be the ultimate format which cannot be superseded? In some ways Digital Betacam does seem already to be very close to being the ultimate format since it answers the lingering doubts over Betacam SP and it offers better quality, arguably, than D1. But there is already a massive investment in Betacam SP worldwide. If Betacam SP is delivering an image which is very satisfactory for most viewers, why change? And where are you going to get rid of all that redundant SP kit? The solution to this problem is backwards compatibility. Certain models in the Digital Betacam range are equipped with SP playback capability. This means that it is possible to replace an SP installation on a planned schedule and not necessarily all at once. The only problem about this is that the more you think about it, the more difficult it is to upgrade gradually. You cannot start by buying just one camcorder to test the water since you can't then edit the tapes, not conveniently anyway. You could start with one camcorder and one studio VTR with SP playback, but then that camcorder's tapes could only be used in that one edit suite. Perhaps it is something more like a confidence-building exercise in the new format, but confidence really is exactly what you need when you make such a momentous decision to change from one format to another.

The kit

Probably the highest profile item in the Digital Betacam catalogue is the Sony DVW-700P, not forgetting the DVW-700WSP widescreen compatible version of course. Some people can remember all these model numbers! The DVW-700P, in both its versions, is a one-piece camcorder, 'one-piece' meaning that the camera and recorder do not separate as is the case with some Betacam SP models. The camera incorporates three $\frac{2}{3}$" CCD sensors (which for some reason are never specified in millimetres) with 1038 horizontal × 584 vertical pixels leading to 10 bit analogue-to-digital conversion and subsequent 14 bit processing. Since 10 bits can capture a wider dynamic range than can be displayed on a typical TV receiver, there is the option to create various 'looks' according to the requirements of the production. The gamma curve is digitally adjustable and is used to balance the importance of the bright and dark components of the scene against the desired mid-tone contrast. Knee correction reduces contrast in the highlight regions to extend the dynamic range further. Variable detail enhancement varies the horizontal detail frequency over the range 2.0 MHz to 6.5 MHz to allow the frequency of the detail enhancement to match the content of the scene. Skin Tone Detail is an intelligent function which reduces the level of detail in skin tone areas of the picture, resulting in freedom from facial blemishes while preserving detail in other areas of the scene(!). I could go on to mention other features of the DVW-700, but I think you get the idea. Almost all of these functions can be controlled remotely, which is essential for multi-camera work. Probably most importantly, once the desired look has been achieved, it can be stored on a memory card. The memory card can be used to build up a library of looks, or it can be used to set several cameras to the same look, which is a useful convenience.

The accessories

I mentioned the look of film earlier. Maybe what is more important is the look of the camera. Perhaps one of the key differences between shooting on film and shooting on video is attitude. And if the camera looks like a film camera then perhaps you will slip into film mode almost instinctively. Sony have encouraged film-orientated manufacturers to provide a range of film style accessories which would encourage existing film camera operators (and film Directors of Photography) to consider Digital Cinematography. These include matte boxes, follow-focus accessories and camera supports which allow the almost painless replacement of a film camera with a Digital Betacam camcorder. In addition the overnight processing of film rushes is eliminated since video tapes are, obviously, ready to play back immediately. For similar reasons video assist, as provided on some film cameras for the director to view the image as seen by the lens, is totally unnecessary.

Figure 6.6 Sony DVW-700 Digital Betacam camcorder.

Whether this really is the beginning of the end for 16 mm and Super 16 film remains to be seen. Opinions vary between those who say that 16 mm should have been strangled at birth and nothing less than 35 mm is good enough for any purpose, including TV, to those who swear by the look of film and would still use 16 mm however wonderful video could possibly be. In between there is a large camp of waverers who can see advantages both ways. Cost may be the deciding factor since Digital Betacam tapes are very much less expensive than film stock and processing.

The alternative

It is always useful to get an opposing point of view. This particular point was made very effectively by Kodak and it is still valid. The second page speaks for itself. Kodak believe that using Eastman 7245 stock a 15-year-old film camera can perform to the standards of high definition video systems, and keep its competitive edge as film stocks continue to improve.

Figure 6.7 Tempted?

Shooting on Eastman EXR film 7245, this 15-year old camera exceeds any of the High Definition Television standards currently being proposed.

Figure 6.8 Film can still hold its own against current video standards.

Digital Betacam format specification

Tape width	12.65 mm (1.2 inch)
Material	Metal particle
Tape thickness	14 μm
Maximum recording time	40 minutes (camcorder)
	124 minutes (studio VTR)
Tape speed	96.7 mm/s
Recorded data rate	125.58 Mbps
Shortest wavelength	0.587 μm
Track pitch	26 μm
Helical tracks per field	6
Longitudinal tracks	Timecode, Control, Audio cue

Video

Input signal	Y, R – Y, B – Y component
Sampling frequency	Y: 13.5 MHz
	R – Y, B – Y: 6.75 MHz
Quantization	10 bits
	8 bits for analogue component
	input of studio VTRs

Audio

Sampling frequency	48 kHz
Quantization	20 bits (18 bits A/D, 20 bits D/A)
Number of channels	4

CHAPTER 7

Standards conversion

Incompatibilities are the spice of life. That is not a phrase to be found in any dictionary of quotations, and it's not a phrase that people who are troubled by the incompatibilities of the many and varied video and television systems in existence would agree with. But as audio onlookers we can take some innocent pleasure in observing the misfortunes of video workers, and perhaps we can consider our own burgeoning issues of incompatibility as we willingly accept format upon format with seemingly no limit to our appetite. Those of us who work alongside film and video will already be coping with as many as six different timecode formats, and we may have to pull up or pull down our sampling rates to accommodate the demands of video, to no good audio purpose. I would guess that any issues of incompatibility that we have now are just the tip of the iceberg, and that standards conversion of audio and video will come to be a major growth industry.

Three problems

In video there are three main problem areas where standards may differ from one system to another. The first is the way in which the colour information is encoded. The three main analogue television systems around the world are NTSC, PAL and SECAM. All of these take a component signal consisting of full bandwidth red, green and blue images, and convert it to luminance (brightness) and chrominance (colour) signals. To economize on transmission bandwidth the amount of detail included in the chrominance signal is reduced. This matches the eye's relative responses to detail in brightness and colour. The first of the systems to be developed was NTSC, which relies on a colour subcarrier being included in the video waveform that encodes the colour information as a phase change between the carrier and a reference signal called the colour burst. Unfortunately this system is sensitive to phase errors in the transmission path and the colours may not always be totally correct. The PAL system was developed later and differs from NTSC in that the phase of the chrominance signal is reversed every line, so that any errors average out. SECAM is completely different and holds the colour signal from one line in a memory and applies it to the next.

The second problem of standards conversion is the structure of the image. In the USA, as you know, there are 525 lines in every frame but in Europe there are 625 (both PAL and SECAM). The inferior spatial resolution of US TV is, however, compensated by the superior temporal resolution of close to 30 frames per second (actually 29.97), compared with Europe's meagre 25. You will notice that NTSC and PAL could hardly be more different – they differ in every parameter. Conversion between PAL and SECAM is more straightforward since they only differ in the method of colour encoding. Their line and frame rates are the same. The third area to which these techniques have relevance is the conversion to and from high definition television (HDTV) standards.

Sampling

Standards conversion is now performed using digital techniques. Whether or not the original video is analogue or digital, during the standards conversion process it will be transformed into the digital domain, and this involves sampling. The sampling of audio involves measuring the voltage of a signal at specific points in time and assigning to each measurement a digital value. This is a simple situation compared with video. Taking things in stages, first of all think of a film camera. This samples an image in time producing twenty-four analogue images every second, which when projected at the same frame rate will give an illusion of smooth motion, most of the time. In analogue video the image is also sampled vertically into lines, although the horizontal element is still a continuously changing analogue voltage. Digital video samples along a third dimension, which is horizontally along the length of the line. The digital image is therefore made up from an array of pixels, sampled from the original scene, which is repeated thirty or twenty-five times every second.

In digital audio, we are all aware that the sampling frequency must be at least twice the highest audio frequency for proper reconstruction of the audio signal to take place. If this is not so, aliasing will occur where new frequencies are created which were not part of the original signal. Obviously, designers of audio equipment pay close attention to sampling theory so that we get pure perfect digital sound, mostly. In film and video, however, sampling theory is thrown out of the window! It is simply not possible for designers to obey all of the rules, and so the design of any system for storing and displaying moving images has to be developed with close attention to the subjective merits of the result. The most frequently seen example of this is the 'wagon wheel' effect, with which aficionados of Western movies will be familiar. This is where the spokes of the wheel may appear to be moving slowly or even turning backwards. This happens because the 24 frames per second sampling frequency is simply not enough to capture the motion with sufficient resolution to fool the eye. We have to accept this because it is impractical to achieve the high frame rate that would otherwise be required.

Even in analogue video, pictures are sampled horizontally and vertically by the CCD array in the camera. In video, frequencies occur spatially as well as temporally. As an easy-to-visualize example of spatial frequency, imagine a row of vertical fence posts: the closer the posts are together, the higher the spatial frequency. Of course in the real world many spatial frequencies combine together to form an image or scene, just as many frequencies combine together to form an audio signal (other than a sine wave). I think it will be apparent that it is quite possible to point a camera at a scene which contains detail at a higher spatial frequency than the CCD array can cope with, and unless preventative measures are taken aliasing will result, forming a pattern in the image which was not present in the scene. Some cameras have an optical anti-aliasing filter which defocuses the image slightly before it reaches the CCD. In interlaced CCD cameras, the output of any one line may be dependent on two lines of pixels which will give a similar effect.

Composite video

Component video, where the three primary colours are handled as separate full bandwidth signals, is great for studios and programme distribution, but it is unsuitable for transmission since it requires three information channels. Composite video only needs a single channel, and in the NTSC and PAL systems a subcarrier which carries two colour difference signals of restricted bandwidth is shoe-horned into the spectrum of a monochrome video signal. To maintain compatibility with black and white television sets (I suppose there are still a few around) the colour subcarrier must not be visible on a monochrome set. The frequency spectrum of a video signal extends up to 5.5 MHz and space must be found somewhere to insert the colour subcarrier. You would not expect to be able to insert an additional signal into existing audio without hearing it, but fortunately in video both the luminance and chrominance signals have gaps in their frequency content at multiples of the line frequency. In NTSC, the subcarrier frequency is carefully calculated to be 227.5 times half the line rate. Because of this, on successive lines the subcarrier becomes inverted in phase. This means that the subcarrier has a two-line sequence, and since the whole frame has an odd number of lines, two frames must go by before the same sequence repeats. This two-frame, or four-field, sequence is an important feature of NTSC which all NTSC video equipment must recognize. There is even a flag in SMPTE timecode which identifies this sequence for editing purposes. In PAL there is a similar sequence, but because the PAL system is slightly more evolved, the sequence becomes eight fields before the pattern repeats. What all of this amounts to is that the luminance and chrominance information becomes interleaved in two dimensions – pretty mixed up in layman's terms – and it is the function of the standards converter to sort it all out.

Interpolation

If you imagine a single pixel taken from a recorded video image, then it will be apparent that it carries meaning in three dimensions. It has width and height, and it persists on the screen for a certain length of time. In digital video it is a single element representing the original image and it appears in its position carrying certain luminance and chrominance values as a result of the NTSC or PAL encoding. But when you see it on the screen, that pixel will be re-created using a continuously varying analogue voltage signal to drive the electron bean that excites the phosphor of the television tube. The original pixels were produced at a rate determined by the digital PAL or NTSC signal, and even though any one pixel comes off the tape at only one point in time, when the signal is eventually converted to the analogue domain, it merges with adjacent pixels to form the analogue signal. In fact, at any time which does not exactly correspond to the instant a sample was taken, the resulting analogue signal is the sum of the contributions of many samples, taken in correct proportion. The result of this is that it is possible to calculate the luminance and chrominance values of 'in between' pixels positioned at points in time where samples were not originally taken. This may sound like guesswork, but engineers call it 'interpolation' because it almost exactly replicates what would have happened had the samples been taken at these in-between points. If all of this is not clear to you, then think of it as a better way of converting a digital NTSC signal into an analogue waveform and resampling it back into PAL. That would work, but the results would be less than optimum.

The simplest form of interpolation takes place when the sampling rate is exactly doubled, as may happen in a very large TV display where lines have to be doubled to make the apparent resolution adequate. This is known as integer ratio interpolation. More complex is the relationship between the 525 lines of NTSC and 625 lines of PAL. But here there is a periodic relationship between the two line rates and there are only twenty-one positions in which an output line can occur between input lines. In this case it is possible to use a system called fractional ratio interpolation which runs at a common multiple of the two rates, so that in-between samples are computed from the NTSC signal, now at the common multiple rate. This data stream will now contain samples at the precise times necessary for the 625 line PAL system. Of course this is very wasteful since many more samples are calculated than will ever be used, and it is possible to carry out this procedure but only calculate the wanted samples. In converters which employ motion compensation or change the aspect ratio, variable interpolation is necessary since there is no fixed relationship between the timing of the input samples and the timing of the output samples. It is not possible to cope with the infinite number of possible relationships, so approximations have to be made which may result in programme modulated noise.

Motion compensation

Neither film nor video systems convey the impression of motion properly because the sampling rate – 24, 25 or 30 fps – is far too low. Film has the problem of judder because each frame is projected twice to prevent the eye detecting the flickering of the image. Anything that moves in the image will not shift evenly as the still pictures are displayed but will judder because it is shown in the same place twice before it moves. This effect may not be obvious, but there would be a perceptible difference if film were shot and projected at 48 fps without repeating frames. In video, if a tube camera and display are used, this judder does not occur because the image is built up line by line over the duration of the frame and shown in the same way. If a CCD camera is used, the image is sampled all at once, like a film camera, but a scanning display builds up the image over a period of time. This results in sloping verticals on moving objects, the degree of the slope depending on their speed. The same effect occurs if a film is telecined onto video since a film frame is sampled at one instant but the telecine scans it over a period of time.

While on the subject of telecine, let us mention another motion-related problem. Film is shot at 24 frames per second and shown on NTSC television at 30 fps. Obviously the film cannot simply be shown at the faster rate since viewers would call the station and complain. The solution is to show one frame of film over two fields and the next over three fields. This is known as 2:3 pulldown. 2:3 pulldown may be just noticeable to viewers but the problem is compounded when a production is shot on film, edited on NTSC video and then transferred to PAL or SECAM for European viewing. Converters have been designed that recognize the third field and discard it, making the conversion easier. The problem with this approach is that when the material is edited on video, the third field will not always end up in the expected place.

It would be possible to convert a 30 fps signal to 25 fps simply by discarding one field in six. With still images this works fine but with moving images there will be jerkiness at a rate of 10 Hz. It is possible to attempt to filter out this 10 Hz component, and this can be achieved by so-called four-field converters which interpolate pixels using data from four fields to compute the likely values they should have. This does eliminate the 10 Hz effect, but motion is still not properly corrected. If an object in the image is moving, then it will be in different places in successive fields. If interpolation is done between a number of fields then this will result in multiple images of the object, and the dominant image will still not move smoothly. If the camera is panned to follow the moving object, then the same effect will happen to the background of the scene. To avoid these undesirable effects, motion-compensated standards conversion works by sensing the boundaries of objects that are in motion and then calculating what are known as motion vectors which trace the position of the object as it moves through the width and height of the picture, and also through time. This is called motion estimation, and the correction employed is properly called motion compensation.

Since video signals do not have an adequate sampling rate as far as time is concerned, motion estimation will not always work properly, and therefore compensation will not always be as good as one would like. An input signal which had a very irregular motion may not be easy to handle. Sporting events, for example, have a wide range of motion, but the eye may be tolerant to slight inaccuracies when a lot of things are happening very quickly. Film weave is another matter as the image is wandering very slightly from side to side, but in a manner in which the eye could very easily notice. Motion-compensating standards converters can be specified on their motion range, sub-pixel accuracy, or the number of different motions that can be handled at once.

It is interesting to consider what is happening when an object is moving over a background. As the object moves, background pixels at the leading edge of the movement will be covered up, while pixels at the trailing edge will be revealed. Therefore when creating any given output field, the converter must look to an input field which took place earlier and apply its leading edge pixels to the output, covering up a section of background. It must also look to an input field which is at a point ahead of the output field in time, to see what trailing edge background pixels are about to be revealed. It seems that an element of clairvoyance must be built into any motion-compensating standards converter!

Comparing standards converters

It would be possible to perform endless technical tests on a standards converter without ever knowing whether it was actually any good or not. The only proper test is a subjective one – do the pictures look right? It is important that the test material is complex enough to tax the abilities of the converter otherwise any shortcomings may not be readily apparent. If a tube camera is used for the source material then there will be quite a lot of motion blur. No standards converter can remove this, but more importantly it may conceal inadequacies. A CCD camera will provide better test material, better still if the camera has an electronic shutter so that every frame is crystal clear. To test the motion compensation abilities of a converter then there has to be some motion in the picture, but note that a camera following a horse on a race track would not be very useful since the camera would be panning to follow the horse. All the real motion in the picture would be in the background and since race tracks have relatively featureless backgrounds the test would be too easy. Ice skating is, apparently, a good test for motion compensation because there is likely to be advertising

Alchemist with Ph.C standards converter

Alchemist is an 'anything to anything' converter capable of very high image quality. Ph.C stands for Phase Correlation motion estimation where a Fourier transform is performed on large areas of the picture. Fourier transforms make it possible to derive the correlation between subsequent fields. Because Ph.C works in the frequency domain rather than the spatial domain it has the advantage of measuring only the movement of the picture, not the picture itself.

Figure 7.1 Snell and Wilcox Alchemist Ph.C standards converter. (Courtesy Snell and Wilcox)

in the background which is clearly visible and in relative motion all the time. Another suitable type of test material which motion compensating converters may find difficult is a scrolling caption, particularly if it is combined with a fade. Stationary captions with moving backgrounds will test the obscuring and revealing processes.

Perhaps the most interesting aspect of standards conversion, I find, is the fact that it can only be judged subjectively. Designing a standards converter which works well is obviously a highly technical task, yet the opportunities for actually measuring the quality of the results are limited.

CHAPTER 8

The video monitor

First, take a look at Figure 8.1. If you think that television is complicated, well yes it is, but I wouldn't say that it is any more difficult to understand than the workings of an analogue mixing console, for instance. Externally on a domestic set the controls are usually very simple, but that simplicity belies all the clever technology that is going on underneath. If I go through the various processes stage by stage then it shouldn't seem so complicated after all; not exactly a painless learning procedure perhaps, but then learning never is. I should clarify the difference between a monitor and a receiver just in case it may cause confusion later on: a monitor takes video signals from a cable input from a camera or other video equipment. A receiver takes an input from an antenna which picks up video signals modulated onto a radio frequency carrier. In other

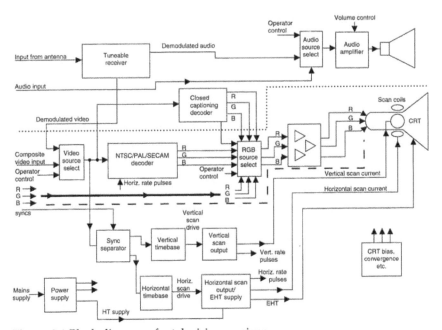

Figure 8.1 Block diagram of a television receiver.

words, the only difference between the two is that a receiver has a device to tune in to whichever broadcast channel you want to watch and convert its modulated waveform into an ordinary video signal. There may be other differences arising from the different ways receivers and monitors are used but these are just details – important details maybe, but at least we should all be sure what we are talking about. In Figure 8.1, you will notice a dotted line. Below this is the circuitry necessary for a video monitor, while above the line is the additional circuitry you would typically find in a receiver. Below the dashed line is the circuitry which, besides the video and audio circuitry, makes the whole thing work. This is where I am going to start my explanation of monitors and receivers, and when you have finished reading this chapter, the whole of Figure 8.1 will be absolutely crystal clear. Really!

Sync and scan

A composite video signal consists of three basic elements: video, audio and sync. The need for a sync signal arises from the way the picture is built up as a sequence of lines and frames, 625 lines per frame and 25 frames per second. The pattern of lines, by the way, is known as a raster. The lines and frames are transmitted sequentially from the TV station and if they are not displayed correctly the picture will 'roll' vertically either up or down. Setting the vertical hold manually used to be a common thing to have to do, particularly when a non-transistorized set had passed its prime. At the end of each line of picture there is a sync pulse which tells the scanning circuitry that it is time to go back and start on a new line. At the end of each field (a field consists of 312.5 of the lines that make up the entire frame) there is a sequence of longer pulses which indicates that the bottom of the field has now been reached and it is time to start back at the top. A further sequence of broad pulses indicates the end of the frame.

Obviously, the video and audio signals can only be a hindrance to any circuitry that is designed to process the sync signals, so the complete video signal – the demodulated video emerging from the video source select block in Figure 8.1 – is passed through a sync separator to strip out the sync pulses from the other elements. The sync pulse, as has been described in Chapter 2, is a 0.3 V negative-going pulse which lasts a mere 5 µs approximately, and of course it was specifically designed to be easily separable. One problem arose when colour information was added to the original monochrome video waveform. In a 100% saturated, 100% amplitude colour signal, the colour information will extend well down into territory that should be the sole domain of the sync pulse. Fortunately, it is a simple matter of adding a resistor and a capacitor to filter out the high frequency chrominance signal before the sync separator so that it doesn't mistrigger. Since the field sync pulses are the same voltage as the line syncs, although longer, they can be separated out at the same time so now we have a signal which has only line, field and frame syncs with no video or audio information.

Now we have successfully separated the sync pulses, we can move on to the horizontal and vertical timebase circuits. The timebase circuit moves the electron beam horizontally and vertically according to the timing given by the sync pulses. In Figure 8.1 you will notice that there are scan coils positioned around the neck of the tube, in what is known as the yoke. These coils attract the electron beam by magnetism and move it vertically and horizontally. Notice that the vertical scan coils are sideways orientated and the horizontal coils run between top and bottom. Is this correct? Yes! When a magnetic field interacts with an electric field (produced by the flow of electrons in the electron beam) the result is that the magnetic field, the current and the motion are mutually at right angles. Both timebase circuits produce a more-or-less sawtooth waveform where the voltages increase steadily in magnitude and then drop back. As the voltage is increasing, the spot is tracing its way at a measured pace over the screen. When the voltage falls, it flies back to its starting position. The simplest type of timebase oscillator will run at a frequency slightly lower than necessary and will be triggered each time by the horizontal or vertical sync pulse. In the case of vertical synchronization this method is good enough but there is a problem with horizontal sync. If the horizontal timebase is driven directly by the sync pulses, its timing will be affected by jitter caused by noise in the signal. For this reason, the horizontal timebase oscillator uses a 'flywheel' circuit which averages out the timing over several tens of milliseconds.

Antenna to CRT

In the upper left-hand corner of Figure 8.1 you will see a box labelled tuneable receiver. There are a lot of processes going on in this box but since it is all to do with radio reception and little to do with video let us just think of it as a box which takes an input from the antenna and produces nice clean video and audio signals. You can see where the audio signal is going so I don't have to explain that. The video signal progresses downward through an input selector. In this case you have the choice of an off-air picture or direct composite video input. From the selector the signal passes to a decoder which could be NTSC, PAL or SECAM. Out of the decoder come three signals, red, green and blue, each with the same relative intensity as the red, green and blue components of the image picked up by the camera in the studio. These signals feed to the three amplifiers which connect directly to the electron guns in the tube. By now the seemingly complex Figure 8.1 should at least be reasonably simple. A colour television or monitor has many building blocks, but each is there for a reason, and it performs its own particular task.

Display technology

Perhaps the most interesting component of the television receiver or monitor is the cathode ray tube itself. There are several basic requirements of any display technology for it to produce an acceptable colour picture on a screen of any kind. It is accepted that to produce a coloured image it is sufficient to build up an illusion of full colour from red, green and blue components, since our eyes only have receptors for these three colours. The first requirement is to be able to position every part of the image on the screen in the same relative position as in the original camera image. The second requirement is to be able to control independently and correctly the intensity of each primary colour at each point on the screen. The first requirement will be met if the display system has correct geometry, which means that straight lines will be straight and circles will be circular and not oval, hopefully even at the screen edges. Also implied in the first requirement is correct convergence. Convergence means that the three colour images should be properly registered. If this is not achieved, objects will be less well defined than they should be and will show spurious coloured edges.

There are two different ways a picture may be built up from the video signal: as a raster scan or as a pixel display. LCD colour displays are pixel devices, as you can clearly see if you look very closely. Some very large displays have been built which use coloured bulbs as the pixel illuminators. In theory, a pixel display could have perfect geometry and convergence, limited only by the problem of addressing the pixels correctly. But the great disadvantage is the number of pixels required, over 300 000 for a 625 line display, which leads to difficulties in manufacture and therefore high cost. Although LCD displays may one day provide the flat screen TV sets everyone has been searching for, they are still too expensive for general domestic use.

An early colour TV system used a monochrome cathode ray tube with rotating coloured filters in front which appeared to modulate the colour of the phosphor in sync with the red, green and blue signals being transmitted. Needless to say this did not work very well, by all accounts, and resulted in pronounced flicker on areas of primary colour, in particular, which would only be shown for one scan in three, and by rainbow-like colours on moving objects. The alternative is to produce red, green and blue images and somehow superimpose them. One way to do this is to have three CRTs, one for each colour, and combine the images in some way. Projection televisions work like this, where each colour drives a high intensity CRT whose images are projected and converged onto a screen. This is inevitably expensive and the images must be carefully aligned. Another less than totally successful scheme is known as beam indexing. In this system there is a tube, the inside of which is coated with a repeating sequence of phosphor stripes which glow red, green and blue in turn when struck by a beam of electrons. This technique offers the possibility of low power consumption or high brightness but requires a high quality electron gun to produce a very narrow beam of electrons, and a very accurate deflection system.

The shadow mask

It sounds like it might have been a good name for an Errol Flynn movie, but actually it is one of the simplest and most brilliant inventions ever. It was developed by RCA in the 1950s and the principle is still in use today in the vast majority of television receivers and monitors. The shadow mask tube has three electron guns sharing a single glass envelope and deflection system, and the three coloured images appear on the same screen without any need for optical recombination. As shown in Figure 8.2, the screen is coated with phosphor dots arranged in groups of three known as triads. The three electron guns produce 'red', 'green' and 'blue' beams, each of which strikes only the phosphor dots of its own colour. I ought to say that the electron beams are not coloured themselves, but their intensities correspond to those of the red, green and blue signals. Before I come onto the shadow mask part, however, let me run quickly through the basics of the rest of the tube.

The electron gun, as you might guess, produces a beam of electrons. It does this by heating up a cathode, which makes electrons 'boil off' to form an electron cloud. The electrons are accelerated towards the screen by a high positive voltage (around 25 000 V) and are focused into a beam electrostatically. Notice that the EHT connection is made through the wall

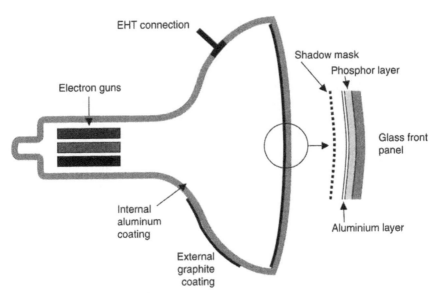

Figure 8.2 Cathode ray tube.

of the tube and not at the electron gun end. This is because the voltage is so high there is a definite risk of arcing. There is also a definite risk of getting your fingers burnt should you ever touch this connector, or the EHT supply, in a TV set or monitor, even if it is switched off.

One of the problems with tubes is that the physical geometry of the tube does not really fit in with what is necessary to achieve a good picture geometry. Consider the spot tracing out a line of picture under the control of a sawtooth timebase signal. It is being deflected from a position in the centre of the yoke around the neck of the tube, so for it to be able to travel at a constant speed across the face of the tube, all parts of the face would have to be equidistant from the point of deflection – the deflection centre. In other words the radius of curvature would have to be quite small. Obviously, we want tubes to be both flatter and squarer. This means that not only is the radius of curvature much greater than would be consistent with such a simple scanning system, but also the curvature is not constant. In practice, the scanning waveform has to be modified, both horizontally and vertically, to avoid what is known as pin-cushion distortion.

Now for the shadow mask itself. As I have said, the beam from the 'red' electron gun must illuminate only the red phosphor, and the same principle applies to the green and blue guns. Figure 8.3 shows how this is achieved. Just before the beam strikes the phosphor dots it encounters a thin metallic sheet with thousands of very precisely aligned holes, around 500 000 in fact. At any point on the screen, each of the three electron beams will strike the shadow mask at a slightly different angle. This makes it possible to screen off the phosphors so that the red beam can only ever strike red phosphors. Now if you think about this, (a) it sounds so simple that you wonder why Benjamin Franklin didn't invent the idea, and (b) it seems almost impossible that a sheet of half million holes could actually be manufactured so precisely in the quantities necessary for large-scale production. In fact, the shadow mask is produced photographically. A steel sheet is covered with a photo-resist material and exposed to the required pattern of dots in ultraviolet light. Where the light strikes the surface, the photo-resist can be developed away, and when the steel is etched the precise pattern of holes is produced. If you are now wondering how this is aligned correctly to the phosphor dots, wonder no more. The shadow mask is used as a template to produce the pattern on the inside of the face of the tube.

Progressive scan

A PAL image, as displayed, consists of a mere 575 visible lines. Most viewers will instinctively sit far enough away from the screen that the line structure becomes imperceptible. The recommended viewing distance is somewhere between 4.5 to 6 times the screen height. At this distance, you

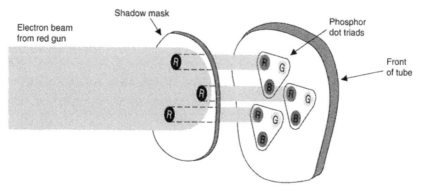

Figure 8.3 The shadow mask.

may not be consciously aware of the line structure but flicker is still a problem. Looking directly at the screen you may not see the 50 Hz flicker, but if you glance away from the screen for a second then the edges of your retina will pick it up. The result is that looking at a TV for any length of time is tiring, and it is something of a mystery that some people can manage to do it for hours on end! The next step in television technology that will reduce both the line and flicker problem is progressive scan. The idea is that rather than interlace the two fields of each frame so that half of the lines are displayed leaving gaps to be filled in by the remaining lines from the second field, all the lines of each frame are digitally stored so that they can be read from the memory in sequence, so the electron beam travels just once from the top to the bottom of the screen to display a complete frame. Since, due in part to digital image storage, it can do this in the time normally taken up by a single field, there is time left over to display the frame again. This results in an image which is easier on the eye and flickers less. The viewing distance can be decreased to something comparable to the way in which a computer monitor is used (computer monitors have used progressive scan for years).

Obviously, cathode ray tube displays are coming to the end of their usefulness as other technologies develop. Even so, it is a demonstration of the power of technology to provide continuous improvements even when ultimately limited by the NTSC and PAL standards developed in the 1950s and 1960s. If a viewer from the early days of colour TV could see the sets we have now, they would probably wonder why we would ever need to go to the movies!

Flat panel displays

The technology behind the cathode ray tube dates back more than a hundred years now and, although it certainly can be said to have stood the test of time, is starting to look rather tired. The cathode ray tube or

Trinitron

Since 1968 virtually all Sony television receivers and monitors have used their proprietary Trinitron cathode ray tube rather than the RCA shadow mask. In the Trinitron tube the shadow mask is replaced by an aperture grille with slots rather than holes. A single electron gun with three cathodes projects three beams towards the screen in the same horizontal plane which are deflected by an electronic prism towards the slots in the aperture grille. The advantages of the Trinitron tube are that the image can be sharper, at least than earlier shadow mask designs, and that the screen is only curved horizontally – vertically it is flat. The disadvantage is that the aperture grille has to be supported by fine horizontal wires which are sometimes visible and interpreted by the viewer as a fault.

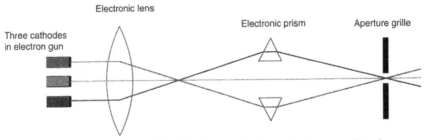

Figure 8.4 Optical analogy of the Trinitron cathode ray tube. In reality the prisms are electronic.

CRT has developed to the point where it can display bright, detailed colourful pictures to the satisfaction of everyone including the most demanding graphic artists and image manipulators. But at the end of the twentieth century it had nowhere left to go. Every last shred of potential had been fully exploited and there was no possibility of making any significant improvement – at least, any such improvement would come at an unsupportable cost. So now we turn on our old friend and pay attention to the less attractive side of its character and although the images we see in the cathode ray tube may be of excellent quality, in a number of ways, we notice more and more how much space its takes up, how much heat it generates, how much radiation in a variety of undesirable forms emanates from its surface. And, by the way, it's pretty heavy too. The time is ripe therefore to look for technologies which will allow a compact display, taking up very little depth behind its surface area, hopefully running reasonably cool with minimal radiation. Technologies that will allow an image quality comparable with the CRT right now, yet with the potential to surpass its qualities in the relatively near future.

LCD

Liquid Crystal Displays have been with us since the 1970s, starting out as crude devices with limited applications including the then-trendy digital watch. I am sure that when an LCD was used for the first time on a piece of audio equipment it was hailed as a triumph of modern technology. The combination of LCD and a couple of nudge buttons often used to pass for a control interface in equipment of the 1980s and somehow we managed. As technology progressed, LCDs got larger and eventually acquired colour. Now we have high quality, high contrast colour LCD displays in notebook computers and larger scale audio equipment which provide excellent functionality in the display component of the user interface. In fact, liquid crystals have a history going back as far as the CRT, but it took many years to find a useful application.

Liquid crystals were first noted by botanist Friedrich Reinitzer in 1888, although he didn't invent the terminology. Crystals, as we normally know them, are solid and often hard, like diamond. Liquids are anything but hard of course, but even so they can form into a regular crystal-like structure, just like soap bubbles can aggregate together albeit on a rather temporary basis. In 1963, scientists at RCA discovered that the structure of liquid crystals could be altered by an electrical charge in a way that, as it happened, turned out to be very useful to display technology. The early days were difficult with liquid crystal materials tending to be unstable, but once this hurdle had been surmounted then steady progress has been made ever since. The first use of a liquid crystal display was in the Sharp EL-8025 calculator and Sharp have continued to be pioneers in the field.

The two characteristics that make liquid crystals suitable for displays are the previously mentioned sensitivity to electric charge, and the tendency of the crystals to align with each other. The basic structure of a display consists of two plates into which is etched a very fine pattern of parallel grooves along which the liquid crystals will automatically align. The liquid crystal material is sandwiched between the plates, which are twisted so that the grooves lie at 90 degrees to each other. The crystals in between the plates will form into a helix with a quarter of a turn, as shown in Figure 8.5. One of the properties of crystalline structures is that light will follow the structure as it passes through, hence light twists by a quarter of a turn as it passes through one plate and then the other. Now, if the two plates are made of polarizing material (the electromagnetic energy of the light is made to vibrate only in one direction) and the plates are orientated at 90 degrees to each other then the matching 90 degree twist provided by the liquid crystals allows the light to pass through. If the liquid crystals were not there, the opposing polarization of the plates would cut out the light completely, but the twist of the crystals in their natural state allows it to pass. We have an 'on' state. Now we need an 'off' state so we have a functioning display. Applying an electric charge to the

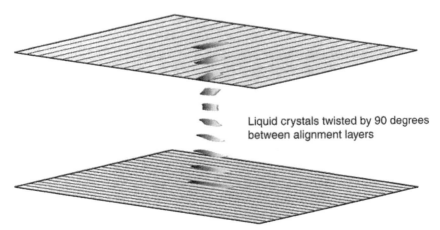

Liquid crystals twisted by 90 degrees between alignment layers

Figure 8.5 Liquid crystals forming spiral alignment.

crystals causes them to disregard their former attraction to the grooves in the plates and line up end to end between them. Now light is allowed to pass between the plates without the 90 degree twist. The opposite polarization of the plates now blanks the display. Fortunately, it is possible to vary the degree of twist by varying the voltage so that in-between grey states are also possible and a greyscale display can be constructed, and colour with appropriate filtration.

One area where flat panel displays differ greatly from the old fashioned CRT is that the energy source in a CRT is an electron gun, or rather one each for the red, green and blue beams. Flat panel displays are constructed from pixels, where each pixel is pretty much self-contained requiring only to be driven with the appropriate signal. Separate pixel displays offer certain advantages over CRTs, and of course have disadvantages as you would expect. The great advantage of a pixel display is that its geometry and focus can be perfect. If you ever get a chance to see the test card on television (which won't be very often these days – and what happened to all that lovely music?), you will notice that one of the functions of the test card is to test geometry and focus. Most CRTs are rather less than perfect in these respects since the distance from the electron gun to the phosphor varies drastically from the centre point to the edges and corners of the screen. This makes it difficult to keep squares square and circles circular. On a rather elderly TV set I keep in a spare room, regular shapes turn into amorphous blobs near the edge of the screen – the complex scanning waveform required to keep things reasonably correct has evidently deteriorated somewhat. Also, you will notice that the definition of the image is rather woolly at the edges, showing up inaccuracies in the focus of the electron beam. The flatter and squarer the tube, the more difficult these problems are to solve. Flat panel displays, on the other hand, pass the test card test with ease since geometry is fixed during manufacture and there is no requirement to focus anything. And there is no possibility of either of these deteriorating over time as there certainly is with CRTs.

Flat panel displays do have their problems, however, and as a fairly new technology this must be expected. Firstly, all pixel displays require each individual pixel to be addressed separately. In a so-called passive matrix LCD this is a relatively simple matter of arranging a grid of electrodes so that each LCD element will lie on a crosspoint. Apply a voltage to the crosspoint and the LCD will become active. This wiring arrangement may be simple, but the drawback is that each pixel can only be activated once in a while, and has to wait while all the other pixels are individually addressed. The liquid crystal material therefore has to be slow-acting so that it retains its data over a period of time until its turn to be refreshed comes round again. This results in a slow-acting display, particularly noticeable on a notebook computer where the cursor 'submarines' if you try to move it too quickly. Some streaking is also evident and is an intrinsic part of the passive matrix technology. To combat this, active matrix (sometimes known as TFT for Thin Film Transistor) LCDs have been developed where each pixel is activated by a transistor and individually addressed. This results in a much faster display suitable for moving images, at the cost of very much greater complexity. The drawback of the extra complexity is that there is more to go wrong in manufacture and defects are inevitable. If only perfect active matrix screens were allowed off the production line then the cost would be astronomical. Anyone who buys an active matrix display will have to tolerate stuck pixels, either in their dark or light state. The manufacturer or reseller will not give you a refund.

One of the other characteristics of LCD displays, passive or active, is that the angle of view is quite narrow. For many purposes this is a disadvantage – it certainly would be if you wanted to make an LCD television, but for private work on a computer in crowded surroundings it turns a drawback into a blessing. The reason why the angle of view is narrow is that the LCD does not emit light of itself, it only transmits it from a light source behind the screen. Head on this is fine, but at even quite a small angle, parallax between the pixels first degrades the image then blocks the view. Making the display thinner obviously helps, but there is a limit to how far this can go.

Plasma displays

One of the hot topics of the moment is the plasma display, currently featuring in television sets on show at the more expensive kind of electrical goods shop. Plasma displays fulfil the requirement of being thin and flat, and the image quality can be extremely good for television purposes, so similar to the standard of a good conventional CRT TV that if you look behind the unit you will wonder where all the works have gone. Plasma displays have individual pixels like LCDs, but there the similarity ends. A grid of electrodes addresses the individual picture elements and applies a high voltage to a gas at low pressure. This generates invisible ultraviolet light which strikes a phosphor. Subpixels arranged in triads glow red, green or blue in proportion according to the requirements of the video signal. If this sounds familiar then doubtless you are remembering the principle of operation of the fluorescent lamp which you probably learnt at school. It is the same, just smaller – and adapted to the requirements of display. Such an established technology has obviously been around in displays for some time, but plasma displays have traditionally been rather low in contrast resulting from a need to maintain pixels in a slightly on state all the time so they can react quickly and come to full brightness as and when they need to. The pace of progress, however, has brought plasma displays to a very good image quality, with the significant advantage over LCDs that light is emitted from a point very close to the surface and therefore the angle of view is wide.

Another area where plasma displays have not fully achieved in the past is definition. In fact, definition of a sufficient standard for plasma displays to be used for computing applications is still difficult to achieve. The manufacturing process does not allow for reliable production of pixels under 0.3 mm across, and the highest definition is only achieved using Fujitsu's ALiS technology (Figure 8.6) which uses an interlaced scan. Progressive scan has been standard in computers for years and is becoming of significance in the TV market too. ALiS (Alternate Lighting of Surfaces) displays cleverly allow two scanning lines to be controlled by three electrodes, rather than the normal two electrodes per line, thus increasing the resolution and eliminating the black stripe between lines that you may otherwise notice. Despite these problems, manufacture of large plasma displays is an achievable aim. Where LCDs inevitably

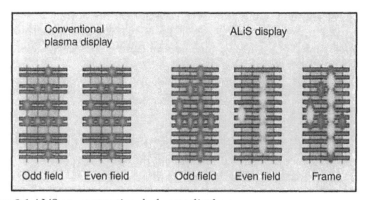

Figure 8.6 ALiS vs. conventional plasma display.

Figure 8.7 Fujitsu plasma display. (Courtesy of Fujitsu General (UK) Co. Ltd.)

acquire more defects as they get larger, plasma displays are currently available with a 42 inch diagonal. Put that in your living room and it certainly will be the centre of attention. What's more, it is perfectly possible to have a thousand or more lines, which makes plasma technology ideal for the HDTV broadcasts currently happening in the United States, of varying standards, and oft mooted in Europe.

Future technologies

The field of flat panel displays is, it seems, up for grabs and a number of companies are engaged in development of proprietary technologies unrelated to LCD and plasma. One such is the Field Emission Display or FED which advances cathode ray technology into the twenty-first century. In a conventional CRT there are three electron guns, one for each primary colour. This configuration is responsible for all of the CRT's major problems – its size, geometry and focus. If the positive aspects of the CRT could be retained while reconfiguring it as a pixel display then perhaps we would have the ideal solution. Basically, therefore, the FED consists of thousands of pixels, each of which in itself is a miniature cathode ray tube complete with electron gun and phosphor. The phosphors, by the way, are standard and therefore can have the excellent brightness and purity of colour of a good conventional CRT. One component of the conventional CRT that is not necessary is the shadow mask, normally used to ensure that the electron beam from the 'blue' gun strikes only the blue phosphor, etc. The shadow mask might be a brilliant invention (as is the equivalent aperture grille in a Trinitron tube) but it wastes a lot of energy. FEDs can therefore be more efficient, as will be appropriate to applications in portable equipment. Other advantages include the fact that light comes from close to the surface of the display therefore the angle of view is wide, and also that the pixels only consume energy when they are on, unlike LCDs where the backlight is on all the time.

Another promising future technology is the Light-Emitting Polymer or LEP. These consist of so-called conjugated polymers as are already used in transparent conductive coatings, battery electrodes and capacitor electrolytes. Certain of these conjugated polymers can be made to emit light in a similar manner to a light-emitting diode. The idea is simple, the materials are there (they have been around for over a hundred years) and all it required was intensive research effort to refine materials that would improve output efficiency and cover the full spectrum of colours. They can be addressed in a similar manner to LCDs (which of course raises the question of whether they will suffer from stuck pixels) and, interestingly, they can be curved and possibly even flexible. The potential uses of curved or flexible displays requires a little imagination, but if they can be made then they certainly will find applications.

Perhaps one of the most fascinating of the emerging display technologies, although not strictly a flat panel device, is the Digital Light Processor, which incorporates a mechanical element in addition to space-age technology (which is now appearing perhaps rather passé). Tiny capacitors are constructed in a silicon surface over which a mirror finished insulating layer is formed. This layer is etched into hinged squares. The charge in the capacitor attracts one of the corners of the square thus changing the angle of the mirror. By reflecting a light from this surface an image can be formed. This component of the DLP is very small, around the size of a postage stamp, and therefore a lot of light is required to form a larger image optically. In some products the imaging chip and its attendant ventilation are mounted in a soundproof enclosure. For colour, three systems are required – one for each of the primary colours – or the chip can be scanned sequentially with red, green and blue light, although this last technique causes the colour of moving edges to break up.

All of these devices relate directly to our audio world, particularly when the way in which we interact with our equipment is a hot topic of debate. There are already large-scale mixing consoles which incorporate several high resolution displays, and of course CRTs abound in any audio-for-video suite. In future, CRTs will be replaced by plasma displays for televisual images, LCDs as we now know them will prosper for some time but ultimately we will see displays and display configurations designed to suit precise requirements of audio. In particular, computer-related audio will no longer seem quite so 'beige box' orientated when the display does what we want it to do, not what the requirements of century-old technology dictate that it must. One last point: none of the specifications of any of these display technologies can easily be related to what they do to single coil electric guitar pickups at close range. Rather an important point to consider before purchase perhaps?

Is your display faulty?

All active matrix displays have faults due to the difficulty of manufacture and vanishingly low yield of perfect devices. Manufacturers quote certain specifications for pixel defects that a customer must tolerate. Devices, according to one supplier, which fall within these parameters are considered good and serviceable: subpixels which are stuck in their bright state must be fewer than fifteen in number, of which not more than six can be green. There may be no more than two subpixel defects within any circle of 5 mm diameter. There may be a maximum of five interconnected defects. The rules for dark subpixels are slightly different: a maximum of twenty, no more than two in any 5 mm circle, no more than two interconnected defects. Most people don't even notice a few defects, and none will ever affect all three subpixels in a pixel. But once you have noticed where they are on your expensive top-of-the-range 500 MHz notebook you will be unhappy. Very unhappy.

CHAPTER 9

Home cinema

Home cinema is a peculiar market segment that stands alone from the mainstream of film, video and audio. Films, naturally, are made to be shown in the cinema, and it is always the producer's aim to recoup the cost of making the film and go into profit from box office receipts alone. Video cassette hire or sell-through and TV broadcast fees are seen as a bonus (sometimes a career-saving bonus). TV programmes are, equally naturally, aimed at a domestic audience who would typically have a large mono audio TV in the living room and a small portable with a correspondingly small speaker elsewhere in the house. So no-one, or virtually no-one, is actually producing material directly for the home cinema market. The home cinema enthusiast will either view laserdiscs or VHS cassettes of feature films, or wait for the still very occasional TV programme made with a Dolby Surround sound track.

The ultimate goal of every home cinema enthusiast is to recreate the spectacle of a film shown at the Odeon Leicester Square in the privacy of one's home. For this you need certain essentials. The first is a suitable space to dedicate to your home cinema. A living room could be perfectly adequate, but for most people this room has to fulfil a variety of functions, with which home cinema may or may not be compatible. One of the ancient tenets of home design is that the furniture should centre around the fireplace, and even though probably the majority of homes now have central heating this does very often seem to be the preferred arrangement. With home cinema, the role of the fireplace is replaced by a large screen television, perhaps a projection TV. When you think about it, where once people might have sat and watched the dancing flames of an evening, now they can watch a dancing electron beam! If you can find space – and the agreement of your co-residents – for a large TV, then you will need large loudspeakers to fulfil the audio side of the equation. I have noticed that audio professionals tend not to be particularly interested in the quality of their home audio systems, or how they are set up. This is probably because we get most of the audio we need in our lives at work, and we are quite content with a small system at home, as long as it doesn't distort and the speakers are in phase of course. For home cinema, however, the enthusiast will want an impressive sound which demands large loudspeakers. Also the left and right speakers will need to be optimally placed either side of the TV – something that just doesn't happen in the average home. Beyond this, real cinema has surround sound in the Dolby, Dolby SR, Dolby Digital, Sony Dynamic Digital Sound or Digital Theatre Systems formats. If it is there in the cinema, then it must be catered for at home too. Having two large speakers either side of the set is more than enough for a lot of people, and adding to them the necessary centre and two surround speakers may be too much to bear. Ideally of course the centre speaker should be equal in size and quality to the left and right speakers. In the cinema this is no problem because all three will sit nicely behind the acoustically transparent screen. At home, although a projection TV could be used of the type that has a separate screen, this is likely to be seen as overkill, and the centre speaker must vie for space and attention with the TV itself.

Despite the inconveniences, home cinema is a growing market, and manufacturers are striving to make their products not only look and sound good, but also to make them acceptable in the home environment. This means that even though the top priority of anyone working in film sound must be to make it sound great in the cinema, at least 5% of one's attention must be given to what it will sound like at home. TV drama perhaps should automatically be made with surround sound so that it will sound good on today's mono and stereo receivers, and will be future-compatible too.

The vision

Curiously enough, it seems that the visual aspect of home cinema is less important than the audio. That makes a change doesn't it? Unfortunately it is as yet impractical to produce a picture at home which is, in proportion to the average living room dimensions, as spectacular as the image you see in the real cinema. Browsing through a few catalogues it is easily possible to find a 50 inch widescreen rear projection TV, which is probably getting on for being big enough. But even the best projection TVs do not give as bright an image as a conventional screen. The largest conventional TV size is 37 inches which, for a TV set, is big (and you would have to consider the depth too!), but is it big enough? I think not. Plasma displays are available with screen sizes up to 42 inches. Of course once you get up to this size then the line structure and lack of definition of the picture become noticeable, which is due to the 625 line standard which is really inadequate, and despite the occasional murmuring about high definition television seems set to remain so for the foreseeable future, at least in Europe.

Connected to the television, even in the biggest and best home cinema, you will find a VHS recorder. Even if it offers S-VHS recording, then as far as I know no pre-recorded material is available in this format. VHS quality is moderately tolerable on a small set, but on a large TV its shortcomings are very evident. The sound on a VHS, owing to the FM sound tracks, can be surprisingly good, and while it is not quite of CD standard many people do find it perfectly satisfactory. A stereo VHS machine has of course two channels, so if surround sound is called for we need a little audio trickery. Fortunately, the technology for squeezing four channels of audio onto two tracks and then extracting them again has been available since the mid-1970s in the form of Dolby Stereo, now just known as Dolby. It is not widely appreciated, but once a film has been made with a Dolby soundtrack, then nothing else needs to be done (apart from decoding the Dolby A or Dolby SR noise reduction of course) to that soundtrack to make surround sound possible in the home, given suitable home cinema equipment. The Dolby Surround and Dolby Pro-Logic systems work on replay only to decode the directional information already present in the film's soundtrack into separate left, centre, right and surround channels. It would of course be possible to re-mix a film for the home market, but from a surround point of view it isn't necessary. An increasing number of TV programmes are now transmitted in Dolby Surround, which as an encoding system is like Dolby Stereo without the noise reduction.

Before DVD the best picture source for the home cinema was the Pioneer Laserdisc system which they claim – believably – has a picture quality 60% better than VHS. Like VHS, discs must be encoded in either

Figure 9.1 LaserDisc player compatible with NTSC and PAL discs.

NTSC or PAL according to their intended market, which inevitably means that most discs are NTSC for the USA. Pioneer offer dual system players so that, if your TV can handle an NTSC input, you can have the best of both worlds. Laserdisc has been going for some time now, but modern players offer convenience features such as being able to play both sides without having to turn the disc over manually. They will also play audio CDs and CDV despite the massive technical differences between these systems.

Audio

If your requirements of home cinema are modest, then you might be satisfied with a surround sound TV. Yes, believe it or not you can buy TV sets which claim to provide a surround sound experience from one unit. Toshiba have their Quadryl system and Samsung have '3D Spatial Surround Sound' where the set has rear-mounted speakers and audio processing to offer a more enveloping sound field. To go a step further then Dolby Pro-Logic sets are available with small satellite surround speakers driven by the set itself. Some have removable left and right speakers too. Of course you have to wonder whether this truly is home cinema and although I don't doubt the worth of sets like this I feel the enthusiast would want something bigger and better.

The something that really is bigger and better is THX Home Cinema. As you may know already, the aim of Lucasfilm's THX system in the

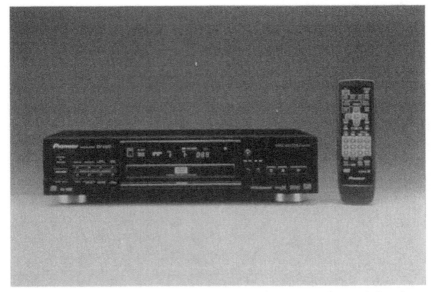

Figure 9.2 DVD player with built-in Dolby and DTS surround sound decoders.

cinema is to make it possible to recreate the sound heard in the dubbing theatre as closely as possible for the audience. THX is compatible with Dolby, Dolby SR and the various digital cinema sound systems and does not compete with them. A THX-certified cinema must use THX-approved equipment and must conform to certain acoustic criteria. These factors must be inspected for initial THX certification and then be re-tested periodically to ensure that the standard is being maintained. While you are unlikely to receive a surprise visit from a Lucasfilm inspector in your home cinema, it certainly is possible to purchase THX approved equipment such as multi-channel amplifiers and special home cinema loudspeakers. THX is compatible with Dolby Surround and Pro-Logic, but whereas other systems may simply decode the channels and amplify them, THX applies enhancements before these signals are reproduced. The first enhancement is 're-equalization', which is intended to compensate for the effect of reproducing sound in a domestic room rather than in a large cinema. A film soundtrack played at home will tend to sound over-bright so high frequencies are reduced in level. Another factor in the home cinema is that the listener is bound to be much closer to the surround speakers than he or she would be in the real cinema, and they will be easily identifiable as point sources. Lucasfilm's 'decorrelation' technique splits the single mono surround channel into two uncorrelated signals for left and right, which will produce a more enveloping sound field. Finally, 'timbre matching' takes into account the difference in the ear's response to sounds coming from the sides and sounds coming from the front. Timbre matching ensures that if a sound moves from front to rear there will be a minimum of change in the quality of that sound.

In addition to these THX enhancements, there are certain technical criteria that THX certified equipment must fulfil. For example, all of

Figure 9.3 The enthusiast's dream, complete with 50 inch rear-projection TV.

the speakers must be capable of producing sound levels of 105 dB SPL without signs of stress. Vertical directivity from the front speakers must be controlled so that the proportion of direct sound is increased in comparison to sound reflected from the ceiling and floor. Horizontal directivity should be wide to cover adequately the whole of the listening audience (all two of them!). Interestingly, the THX specification calls for the surround speakers to have dipole characteristics. This is so that sound will be projected forwards and backwards parallel to the side wall of the living room, and not directly towards the listener. This will reduce the localization of the surround speakers, which is a desirable feature.

How it works – Dolby Surround

The contribution Dolby Laboratories have made to cinema is widely known and respected. Until the coming of digital cinema sound, the Dolby Stereo system was the *de facto* standard for feature films, and even now, a film print carrying a digital sound track will also have a Dolby Stereo track to maintain compatibility with non-digital cinemas. I should point out that Dolby Laboratories now wish Dolby Stereo to be known simply as Dolby when A-type noise reduction is used, and Dolby SR obviously when SR noise reduction is employed. Apparently most people now assume that stereo necessarily means that there are two channels and two loudspeakers where, according to the original usage of the term, that need not be the case. My apologies to Dolby Laboratories but there are so many different Dolby systems now that I am going to have to stick to the Dolby Stereo terminology here otherwise everyone is going to be confused!

The wonderful thing about Dolby Stereo (among several other wonderful things) is that it encodes four channels: left, centre, right and surround (LCRS) onto two optical tracks on the film print. These two tracks can be played as though they were conventional stereo and they will still sound pretty good, so Dolby stereo is compatible with conventional two-channel stereo. If a film soundtrack is encoded into Dolby Stereo, then the four-channel surround sound information will survive all the way through to release on laserdisc or VHS video cassette, and without realizing it many of us invite Dolby encoded sound tracks into our homes on a regular basis and all we need is the right equipment to hear it in its full glory.

Apart from the element of noise reduction, there is no difference between a Dolby Stereo and Dolby Surround sound track. Both contain the same LCRS information. The way in which the encoder works is this:

- Left, right, centre and surround channels must be encoded to two tracks, left and right.
- The left and right channels go directly to the left and right tracks.
- The centre channel is added at –3 dB to both the left and the right tracks.
- The surround channel is band limited between 100 Hz and 7 kHz. It is encoded with a modified form of Dolby B-type noise reduction. Plus and minus 90 degree phase shifts are applied to create two signals which are at 180 degrees with respect to each other. These are added to the left and right tracks.

The original Dolby Surround decoders used a passive matrix to create a phantom centre image from two front speakers, and the surround channel passed through a time delay, 7 kHz low pass filter and a modified B-type decoder. With this type of system, the worst problem could be that signals from the front channels, particularly dialogue from the centre channel, could find its way into the surround speakers. The time delay ensures that any crosstalk is delayed, which will help to fix the image in the front speakers due to the precedence effect. The 7 kHz filter and modified B-type noise reduction help reduce the effects of any remaining crosstalk. The modification in the B-type noise reduction is, by the way, that it offers only 5 dB of processing rather than 10 dB, otherwise the left and right signals could be significantly affected.

Nearly all Dolby Surround decoders currently available are of the Pro Logic design which actively derives a centre channel to keep dialogue firmly fixed in position. A passive decoder without a centre speaker can only work well for listeners in the optimum seating area and the isolation between front and back is not ideal. A Pro Logic decoder uses 'directional enhancement' to eliminate as far as possible the undesirable effects of the matrix system. The key concept is that of signal dominance. At any one time it is likely that one signal will be dominant in the mix and the Pro Logic decoder ensures that this signal will be given directional enhancement while redistributing other signals spatially. There may be a high degree of dominance, perhaps when there is only one signal present, in which case the Pro Logic decoder will apply substantial directional enhancement to make sure this signal comes from the correct point in space. On the other hand, atmospheric sounds such as wind and rain are unlikely to exhibit a high degree of dominance so no directional enhancement need be applied. When dominance is high and rapidly changing, then the decoder will react rapidly. When dominance is low, the decoder will react more slowly to retain stability in the sound field.

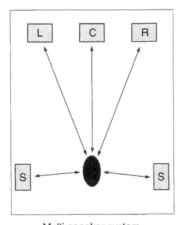

Multi-speaker system

Virtual Surround

Figure 9.4 Room layout for Dolby Surround.

Figure 9.5 Room layout for Dolby Virtual Surround. (Courtesy of Dolby Laboratories Inc.)

Virtual Surround

Of course, we will soon move into the digital surround sound era and Dolby Laboratories' presence will undoubtedly be felt. In the meantime, their latest contribution to home cinema is Virtual Surround sound. It is a well established fact that in any household containing two adults, at least one of them will not be a home cinema enthusiast and will strongly object to ever-increasing numbers of loudspeakers appearing in the living room. Dolby Laboratories' answer to this is to reduce the requirement for five speakers down to just two, as in conventional stereo, although you will have to place them in their optimum positions and not have one on a shelf and the other on the floor behind the sofa!

There are a number of technologies that allow a full 360 degree sound field to be reproduced via two speakers, some more successfully than others of course. Among them are QSound's QSurround and CRL's Sensaura. Dolby Laboratories refer to this type of technology as a 'virtualizer'. Dolby's position in this is to provide additional technology to translate a multi-channel Dolby Stereo or Dolby Digital sound track into a form that the virtualizer can use to recreate a sound field similar to that you might have experienced if you had a 'real' five-speaker system. As long as the virtualizer meets Dolby Laboratories' performance criteria then the result should be effective, although different virtualizers are bound to be dissimilar in areas such as the frequency response of the virtual channels and sweet spot size.

CHAPTER 10

Nonlinear editing

Just as hard disk editing systems have become an intrinsic part of the audio production process, so disks have found applications in film and video editing too. It is customary to call this kind of editing 'nonlinear' in the visual environment, obviously because tape and film store images in a line whereas a hard disk offers almost instantaneous two-dimensional access to any part of the data. Nonlinear editing as a concept really dates back to film, because even though film itself is linear, the 'line' can be split apart and new material cut in – something which video tape cannot do (not since the days of Quadruplex anyway). Conceptually, film is the perfect editing medium since everything that you might want to do in terms of story-telling is equally easy, the corollary of that being that everything is equally difficult. There is nothing to stop you therefore going for exactly the edit you want. In video editing, some things are easy, others are more difficult. Guess what human nature leads to! In the past, film was always seen as a 'fast' medium, since it was more straightforward to try things out and fine tune than on the 'slow' medium of video where there is a distinct tendency to perfect one thing, then another, and another – but never go back. 'Fast' and 'slow' refer to the end result, not necessarily to the time taken in the editing process itself. Another feature of film is that picture and sound have traditionally been handled on separate media, unlike video tape where picture and sound are bonded together. Thus in video tape editing it is normal practice to cut picture and sound at the same place because a 'split' edit is more difficult. On film there is no distinction and edits are done according to what looks best and what sounds best. A film editor would think of that as being entirely obvious.

One of the features of film editing in contrast with video editing is its ability to change anything at any time. In video tape editing it is difficult to change the duration of a scene without a certain amount of panic. It is not possible to insert a shot ten minutes back without rebuilding the entire ten minutes worth of material that follows the alteration, or at the very least copying it. Film editors have never had those limitations. The art of film editing is founded in story-telling and is not driven by work practices necessitated by the machinery. Ask yourself the question: would you be likely to go back those ten minutes to change a shot if you knew that you would have to do all that work over again? Or would you find yourself justifying your original decision? We know what the answer should be, but the temptation to do otherwise is strong. For a long time into the video era it was seen as very desirable that video tape should be as easy to edit as film. Products such as EditDroid and Montage used banks of laser discs and Betamax tapes respectively to allow access to lots of material very quickly, and were seen as ingenious solutions to a difficult mechanical problem of the time. These early systems did, up to a point, allow editing to be done more conveniently.

Nonlinear editing only started really to prosper once computer technology had advanced sufficiently to accommodate real time video images. It is difficult now to think back to the late 1980s and early 1990s when desktop computers were only useful for text, graphics and a little bit of photographic image manipulation (and computer sound editing was in its infancy). Moving pictures take up vastly more data and a means had to be found to balance the quantity of data with the speed of computer processing at the time. Some companies, such as Avid and Lightworks (now absorbed into Tektronix), took the route of reducing the amount of data so that a standard desktop computer could act as the host platform. Others such as Quantel built fiendishly powerful proprietary computing systems that could accommodate a much higher data rate, but at a proportionately higher hardware cost. Fascinating though the high end equipment is, it is the desktop systems that have made an incredible impact on methods of operation in the entire video and broadcasting industry.

The key to reducing the data rate to manageable proportions was JPEG compression. JPEG compression of still images is able to reduce a file size of megabyte proportions down to tens of kilobytes, at subjective image qualities that range from being so good that you can't see any difference from the original, to tiny files that still give a reasonable guide to the original image content, but are not in any sense usable as an end product. The versatility of JPEG is such that once it became possible to achieve JPEG compression of moving images in real time, the application to computer-based editing of video material became obvious.

Offline/online

In the early days of nonlinear editing, image quality was quite poor. This is because the disks and processors of the time could only handle a relatively modest data rate, and the disks were minute in capacity compared with what we take pretty much for granted now – they were expensive too. High end manufacturers rebuilt disks to higher specifications, or made them into 'striped' arrays where the data could be distributed in a way that allowed retrieval of full broadcast quality images in real time. But for the majority of editing applications, the budget just did not allow for the costs these systems entailed. Fortunately, existing working practices allowed nonlinear editing to enter the professional environment very easily. The offline video editing procedure was already established where work would be copied onto a low-cost format and edited in an affordable suite where thinking time was available at a reasonable hourly rate. The product of the offline suite would be an edit decision list (EDL) from which a finished master could be put together from the original source tapes in the much more expensive online suite with full broadcast quality equipment. If a nonlinear system could replace the tape-based offline suite, then it would have a niche from which it would ultimately establish itself as standard working practice.

So it came to be – the process of transferring broadcast quality material onto U-Matic or VHS tapes was replaced by the process of 'digitization' where material is copied onto the nonlinear editing system's hard disk at reduced quality, with the attendant savings in the rate and quantity of data. In fact, the term 'digitization' is still used when Digital Betacam tapes are transferred to hard disk. Strange but true. Once the material is on the disk, then the sequential access of tape is replaced by the random access of disk, both in the selection of material that is to be incorporated into the edited master, and the technology behind the actual playback on screen of the images. In early nonlinear editing systems, standard working practices were modelled quite closely, where appropriate. Thus editors can still think in terms of clips (as in film clips) and bins (where film editors store short lengths of film). In modern terms, the equivalent of a bin would be a folder or directory on the hard disk and a clip would be a file. Avid, for instance, use film rather than computer terminology on the grounds that it is what film and video people understand. This is extended to 'sub clip' which is a section of a clip identified with its own start and end point, although the data for the entire clip still resides on the disk and can be utilized later if need be. Of course, not all parts of the film editing process are modelled. No longer is there any necessity to mark the film with a grease pencil, splice it with tape or put up with the dust and scratches that the work print is inevitably bound to acquire. And you don't have to hold a clip up to the light to see what is on it – a still from each clip, and not necessarily the first frame, can be displayed so that the editor has an immediate overview of the material available. This does actually raise an interesting point – how much information should the editor retain in his or her head, and how much should be presented on the screen? Traditional methods, both film and video tape, have involved reviewing the material and then simply 'knowing' which sections to use, and then assembling and refining the edit. Now it is much easier to devolve that part of the editing process to the computer screen and assemble a sequence from the key frames on display rather than a 'feel' for what the footage contains and the pace of the action. We sound people

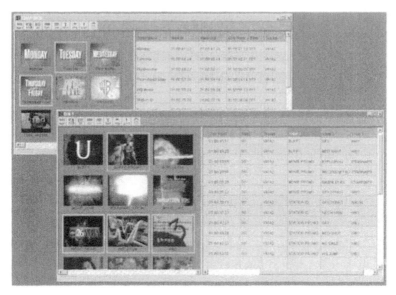

Figure 10.1 Discreet Logic Edit bin window.

of course work by feel all the time, since any visual representation of sound can only be a very crude approximation of the auditory experience.

At the end of the offline process, the product is an EDL, just as it would be in a tape-based offline suite, which can accompany the source reels into the online part of the process.

Timeline

Ever since the early days of audio hard disk editing (always an easier proposition than video) it has been held as axiomatic that a nonlinear editor requires a linear 'timeline' display, since the end product is going to have a beginning, a middle and an end in predefined order. (Of course multi-media is another thing entirely.) This is true of video nonlinear editors too. Typically, a nonlinear editing system will allow clips to be dragged onto the timeline where they can be viewed as a sequence of stills, with time resolution according to the magnification of the display. In the timeline the clips can be topped and tailed as necessary and assembled into the desired order. Just like an audio disk editor, the edit can be played to see roughly how the whole thing fits together. Here we find another interesting comparison with video tape and film. Video tape editing revolves around 'in' points and 'out' points. View a segment of the source reel, decide which frame on which to start the shot and mark it. Now find the frame on which the shot should end and mark that too.

The process involves shuttle, jog and still frame and is very static compared with film where an editor will repeatedly view the film forwards to the edit point, then backwards, forwards again and over and over, continuously refining the edit point in a dynamic sense, always considering the flow of the action. This is impossible with tape, but it is certainly possible with nonlinear editing systems. In fact, some offer controllers that allow film editors who are used to Moviola and Steenbeck machines to feel that they are working in a very traditional way with a traditional film feel. The mouse may be a modern way of working for many applications, but the old film ways have evolved over decades and are unlikely to be bettered easily. (As a comparison, imagine a MIDI composer inputting data into a computer with a mouse rather than a piano-style keyboard.)

Unlike with audio editors, playback of the edit in a finished form is not always possible directly from the timeline. This is because anything more complex than a simple cut edit has to be rendered, which is similar to the way fades are written to disk in some audio editing systems, but probably not as fast. A cut edit is where the outgoing shot ends on one frame and the incoming shot starts on the next. A cut is just the simplest form of transition. Another simple transition is the dissolve, which in audio terms would be called a crossfade. This is where video tape editing systems start to struggle and rocket science has to be applied. Doing a cut edit in a tape-based suite involves copying one shot from the source machine to the destination machine, then simply copying the next shot as a butt edit. Precise timing and automated switching is involved of course but the

Figure 10.2 Discreet Logic Edit trim window.

concept is straightforward. To perform a dissolve, however, involves bringing in a third machine so that there are two source machines (in some cases short dissolves are possible using a 'pre-read' function where available). And what happens if the shots you want to dissolve are on the same source reel? You have to copy off to a 'B reel' and lose a generation. Remember what I said earlier about some things being easy and some things being difficult? Actually a dissolve is even more difficult in film, but the editor doesn't actually have to do it – they just mark it on the work print. The work is done by an optical facilities house. Nonlinear editors can do dissolves quite easily, and other transitions such as pulls, pushes, wipes, etc. Of course, when these transitions are performed offline, they have to be replicated in the online suite, but a selection of standard transitions are commonly available and can be specified in the EDL. Higher-end systems include visual effects such as resize, reposition, flip, flop, luma and chroma key. (A flip, if you were wondering, is like reflecting the image in a mirror. A flop is turning it upside down.)

Integration

The early nonlinear editing systems were definitely offline systems, only venturing into online territory where degraded image quality was acceptable, as in CD-ROM multi-media for example. But as processing speeds increased, and disk storage became cheaper, the quality level got better and better, to the point where broadcast quality is achievable at the right price point for general use. A very mild compression can be used comparable to Digital Betacam, or the 5:1 compression ratio of DVCPRO and DVCAM is indistinguishable from broadcast quality for many people. Once nonlinear editing is available online, then a whole new vista of possibilities opens up. Firstly, the time to air decreases significantly and editing can even commence while material is still being recorded. A typical example of this is a sports match where a replay sequence can be assembled very quickly to drop in whenever there is a lull in the action. In fact, news and sports were the early adopters of online nonlinear editing systems, and continue to pave the way in many respects. Episodic television also benefits greatly from online nonlinear editing since the pace of production is often breakneck. There are stories of extra episodes of soaps being pieced together largely from outtakes that didn't make it into the general run of the programme, and doubtless many readers will have their own experiences of this.

The distinct difficulty with the offline/online process is in reproducing the transitions accurately, captioning and effects, and in fine control over the ultimate appearance of the images. Inevitably this results in a lot of work being done offline, and then there is still a lot of work to do in the online suite. In particular, there might be problems that simply do not come to light in a reduced quality offline system that are only too apparent when seen at online quality. Integrating everything into a single broadcast quality online stage is therefore very desirable, and now possible. Having one decision-making opportunity is significant since all too often minds can change in the lull between the offline and online stages – not always to the improvement of the programme. Consequently, an online quality nonlinear editing system will offer colour correction so that grading from shot to shot can be carried out as a matter of routine. Paint facilities may be included so that the occasional errant boom microphone in shot can be obliterated (and a memo sent to the person responsible if it happens too often). Captions can be created and superimposed on images, and animated with the common rolling and crawling movements. Digital video effects too may be added at this stage, with a degree of sophistication commensurate with the cost of the system, or imported from another system and incorporated into the edit more conveniently than having to remember that there is a problem to attend to at some later time.

In many applications, so-called 'versioning' is of considerable significance, where material might be re-edited into several different forms depending on the nature of the applications. The classic example is news, where reports in different formats are required according to the duration of the newscast, its schedule during the day, and ongoing developments

Figure 10.3 Quantel Editbox. (Courtesy of Quantel.)

in the story. Versioning on tape is obviously not the ideal way of working since work has to start pretty much from the beginning for each new version required, although copying previously edited material may be a useful short cut. Versioning on a nonlinear system, however, accesses the same material on disk simply in different ways. Combining parts of previous edits is absolutely no problem, and there is of course no generation loss. Versioning also exists in promos, where perhaps three versions are required with shots from the programme itself, captions, effects, titles, voice-over (yes, nonlinear editors handle audio too) and – most importantly – attention-getting creativity, which nonlinear systems allow for in abundance.

At the end of the editing process, nonlinear editing systems offer the ability to play out directly to air, which is desirable for news, sports and anything that needs to be shown very shortly after it happened. Direct playout from a standalone workstation leads neatly to the concept of having a central server where material is loaded, accessed and edited, and played out from when necessary. Server-based nonlinear editing has a significant future, particularly in news and sports, but has relevance to episodic production, and any situation where many people are working on the same project at the same time. Once a system such as this is in place, there is no possibility that anyone is likely to want to go back to the 'good old days' of video tape. In retrospect, video tape never really had any good old days and the future of editing is definitely nonlinear – as it was with film. We seem to have completed the circle.

CHAPTER 11

JPEG and MPEG2 image compression

The cost of data storage may have tumbled to a fraction of what it was even five years ago, and the Internet's bandwidth is constantly increasing, but the requirement for storing and transmitting data in ever smaller packages remains. We now have effective audio data reduction technologies, but before that came data reduction and compression of images. JPEG compression for still images is the basis of MPEG, the moving picture equivalent. But before we try to understand image compression, it is probably a good idea to take a look at what an uncompressed image consists of. One way to do this is to think in terms of computer displayed images as it provides a simple starting point, although this is relevant to all digitized image systems. If you look back to an earlier age of computing, particularly to the old compact Macintosh range or Atari ST, you may remember images in which each pixel (picture element) was described by a single bit of data. That 1 bit could be either 1 or 0 and the pixel therefore could only be either fully black or fully white. These screens did not show any levels of grey, except by mixing areas of black and white pixels. When you upgraded your 1 bit monochrome computer to a more modern colour-capable model your dealer would have informed you that the monitor was capable of showing 16 colours, 256 colours, 'thousands' of colours or 'millions' of colours. Colour monitors are in fact capable of any of these; it is the amount of RAM that is dedicated to video inside the computer that is the limiting factor. If you want each pixel to be capable of 16 colours then fairly obviously 4 bits of RAM per pixel are required, multiplied by the number of pixels in the screen. 256 colours require 8 bits, 32 768 colours (thousands) require 15 and 16 777 216 (millions) require 24. 'Millions of colours' is becoming the standard these days and it is easier to think of it as 8 bits for each of the primary colours. This explains why 15 bits are used for 32 768 colours rather than 16 for 65 536; 5 bits each are used for the red, blue and green primaries. You may be asking at this stage whether it is really necessary to have millions of colours. Can the eye really distinguish this many? For most intents and purposes thousands of colours on a computer screen are indistinguishable from millions, but the extra colours are used as 'professional headroom' just as we like to have extra bits – when we can get them – above and beyond the 16 bit domestic CD standard. Even at the thousands of colours standard, a 640 × 480 pixel image demands more than half a megabyte of data. A moving image which consumed data storage at this rate would require a data rate of well over 100 megabits per second. This is impractical for any consumer medium.

JPEG

Of course you have heard of JPEG already, and MPEG too, so let us get the explanation of the abbreviations out of the way. They stand for Joint Photographic Experts Group and Motion Picture Experts Group respectively. JPEG is for still images and MPEG for moving images, but as we shall see shortly there is a class of moving images for which JPEG compression has relevance. As you will agree, while it is important to have standards on compression technologies, it is just as important that the standards that are agreed are the best ones, and not the ones which the most politically powerful manufacturer has been able to bulldozer through. The first and most obvious question to ask is just who are these so-called photographic experts? The simple answer is a working party set up by ISO, the International Standards Organization, and CCITT (International Telegraph and Telephone Consultative Committee, translated from the French). These experts examined how the eye perceives images, and then they proceeded to look for ways in which data could be stored more efficiently, and then they went on to consider how parts of the image that the eye would not miss could be discarded.

We are used to dealing with audio frequencies which describe the number of cycles of a waveform that occur in a given time. There are such things as spatial frequencies, too, which describe how many contrast changes occur in a given angle, with no reference to time. The eye is more sensitive to some spatial frequencies than others, which probably relates to some survival mechanism we evolved to cope with a harsh prehistoric existence on the plains of Africa. It turns out that the peak in contrast sensitivity comes at about 5 cycles, or intensity changes, per degree and drops to zero at about 100 cycles/degree. The former value would relate to viewing objects 2 mm in size at a distance of 1 m. At the same viewing distance we would have difficulty resolving objects smaller than 0.1 mm. Oddly enough, although the horizontal and vertical resolutions of the eye are similar, the response to changes of intensity along diagonals is reduced. The JPEG image compression system takes advantage of this.

Now that we have an idea of the eye's spatial frequency response, we need to know something about the number of steps needed to make a smooth transition from black to white – the number of quantizing levels in an ideal situation, if you like. Under perfect conditions, the eye can distinguish about 1000 levels of grey, which would require about 10 bits per pixel. In practical systems, however, other considerations make it possible to describe a pixel as fully as the eye needs in 8 bits.

Because colour vision depends on three sets of receptors in the retina, compared with one set for luminance (brightness), the resolution is naturally rather less. The peak sensitivity of the eye for chrominance changes comes at 1 cycle/degree (a 10 mm object at 1 m viewing distance), and drops to zero about 12 cycles/degree. As a result of this, it is possible to compress the colour information to a fraction of the data required for the brightness. For this reason, JPEG works best in systems where the image is converted from RGB to luminance and chrominance, such as video and TV. It should be noted that JPEG consists of a set of strategies for encoding and decoding an image. Here I am describing just a subset of the processes involved.

Discrete cosine function

No, don't turn the page! Just because this looks a little bit mathematical doesn't mean that it can't be understood in qualitative terms. The discrete cosine transform (DCT) is one of the basic building blocks of JPEG, providing efficient lossy compression. The DCT is not all that far removed from Fourier analysis, which of course we are all thoroughly familiar with (qualitatively, remember). The process of Fourier analysis (named for Napoleon Bonaparte's top man in Egypt incidentally) separates a complex waveform, which may be an audio waveform, into its constituent frequencies. It is common knowledge now that any audio waveform may be reconstituted by building it up from sine waves of the correct amplitudes and phase relationships. DCT does virtually the same thing; a section of an image can be looked upon as a complex spatial wave and it is converted to its constituent spatial frequencies. When this is done, you will find that some of the frequency components are out of the range of the eye's ability to see. Aha! All we have to do is throw them away and we will have saved some data at no effective cost. The image will look exactly as it did when it is put back together again. In addition to that it is possible to quantize the frequencies so that, for the purposes of explanation, spatial frequencies of 5.25, 6.1, 7.3, 7.9 cycles/degree could be rounded to 5, 6, 7 and 8 cycles/per degree. Of course the real life situation is more complex, but I'm sure you get the idea.

Now you have the basic idea, let me explain DCT with a little more precision: JPEG uses a two-dimensional DCT. A one-dimensional DCT would consist of, say, eight cosine waveforms which are called basis functions. (A cosine waveform is the same as a sine wave but it starts from the position of the maximum value rather than zero.) These waveforms would be d.c. and seven alternating waveforms at particular frequencies. All of these would be sampled at eight points. Any arbitrary

Input 8x8 pixel block

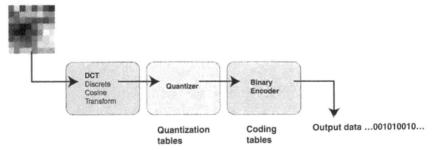

Figure 11.1 Overview of JPEG encoding process.

sampled waveform can now be replicated by combining the basis functions. The two-dimensional DCT is made by multiplying horizontally orientated one-dimensional basis functions by a vertically orientated set of the same functions. This will give sixty-four square patterns each containing sixty-four cells in turn containing black, white or in-between levels of grey (Figure 11.2). They all have a checkerboard-like appearance but are subtly different, covering a wide range of the patterns that can be achieved. These can be combined together in any proportions to produce

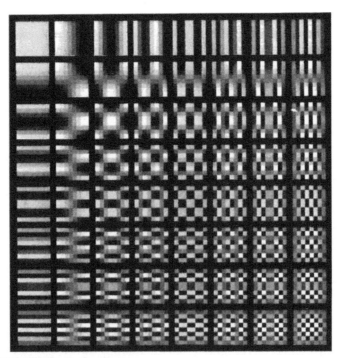

Figure 11.2 DCT basis functions.

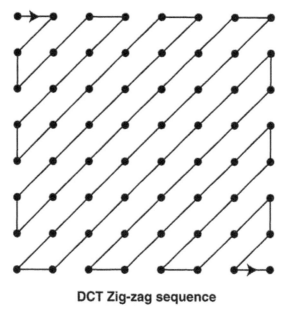

DCT Zig-zag sequence

Figure 11.3 Zig-zag path orders the basis functions from low to high spatial frequency.

any arbitrary grid of sixty-four greyscale cells. The proportion in which each basis function is applied is called its coefficient. It turns out that the eye is sensitive to each of the basis functions to a different degree and it is possible to measure the threshold of visibility for each. The coefficients are then divided by that value and rounded to integer values. At this stage the image can be almost completely restored, but it is also possible to divide by a value larger than the visibility threshold, which will give data reduction at the expense of visible artefacts in the reconstructed image.

I have been talking about 8×8 blocks, and indeed JPEG does work in this block size. However, the size of these blocks can be larger or smaller depending on the trade-off you have decided between image quality and data quantity. One obvious problem that splitting the image up into square blocks causes is that the edges of the blocks become visible. This is the first thing you will notice in a heavily JPEG compressed image.

Entropy coding

Altogether different from DCT and another part of JPEG is entropy coding. The concept of entropy comes from classical thermodynamics and is normally taken to mean the amount of 'disorder' in a system. For instance, there are a lot of hot spots such as stars in the Universe, and a lot of pretty chilly cold bodies, dust and empty space. Eventually all the heat from the hot objects will have been lost to the cooler objects and everything will be at the same temperature. We will then say that the entropy has increased. Actually we won't, and I'll direct you to the physics shelf of your local bookstore to find out why. Entropy coding is a 'difficult' name for something that is pretty simple. It means that things that happen often should be described briefly because we already know what to expect, and that things that happen infrequently should be described at length because they are unusual. That way, we can get the maximum amount of information with the minimum amount of description. Entropy coding involves no loss of information at all. Let me give an example. Suppose that we want to transmit a regular traffic report, and over the course of a year we want to transmit the information with as few bits as possible. For the purposes of this example there are only four types of traffic conditions which we could describe in a binary code:

Congested	00
Heavy	01
Moderate	10
Light	11

This might seem already economical with data but the situation changes if you learn from past records that on average it is congested half the time, heavy one-quarter of the time, moderate one-eighth and light one-eighth. Knowing this, you will realize that it would make sense to reduce the number of bits required to say that it is congested, even if it takes more bits to say that it is moderate or light. In fact, rather than taking up 2 bits for each report, it is possible to use only 1.75 bits per report on average if you use the following code:

Congested	0
Heavy	10
Moderate	110
Light	111

You may say that there is still some redundant information here, but actually the coding (this is an example of Huffman coding) is rather more sophisticated than it appears since the probability of a 3-bit word being 111 is 1 in 8, or one-eighth – exactly the probability that the traffic is light. From my example I think you can see that the principle can be extended to any sort of data where some combinations occur more frequently than others. Huffman code is one type of code used by JPEG, though JPEG has other strategies up its sleeve to minimize the data used.

Results

The quality of JPEG compressed images is visually very good, as can be seen from my example. One point that you may find unusual about image compression is that you do not specify a compression ratio from the outset as you do for audio. Since the reduction that JPEG can achieve depends largely on the content of the image, you will only be able to calculate the ratio after the compressed image file has been produced. The software I use offers a scale with which the trade-off can be set between the amount of compression and final image quality.

Figure 11.4 Original Photo-CD image.

Figure 11.5 10:1 JPEG compression.

The original image comes from a Photo-CD. The image shown here is only a small part of the original 35 mm negative but on the screen of my computer it looks pretty good technically, and it should in print too. Figure 11.4 is uncompressed and just as it came from the Photo-CD. The file size is a little over 2 Megabytes. I tried some low ratio JPEG compressions but the result was so nearly indistinguishable from the original that I haven't shown them here. This is correct of course, because JPEG's first action is always to change the image as little as possible and most likely only

Figure 11.6 20:1 JPEG compression shows only slight degradation which is just visible as 'blockiness' in the background.

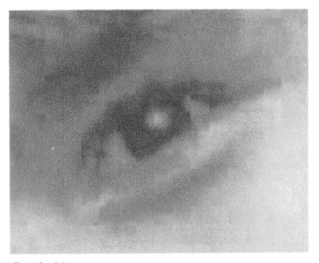

Figure 11.7 Detail of 20:1 compression.

lossless data reduction occurred. Figure 11.5 is the same image but this time compressed down to 100 kbytes, a 20:1 ratio. On the printed page I doubt if you will see any loss in quality, but I can see it if I look carefully at the computer screen. Figure 11.6 is the maximum compression I was able to apply, down to 43 kbytes. You should be able to see the difference now, but in case you can't because of limitations in the printing process I have blown up a small area (Figure 11.7). Here you should clearly be able to see the edges of the blocks created in the discrete cosine transform process. Here we are well into the region of lossy data reduction.

JPEG for moving pictures

I mentioned MPEG earlier, which is a development of JPEG for moving images. Further data reduction is possible because any frame is likely to have a lot of similarities to its neighbours, and therefore there will be some redundant data. Although MPEG is useful for feature films and anything that will be shown continuously, it is not so practical for any application where you might want instant access to any single frame. If MPEG had been used then you could easily want to have access to a frame that was not fully described, so the system will have to backtrack and build the image up. This would take too much time so JPEG coding is used. If anything, JPEG coding is even more impressive on a moving image because the eye tends to average out a number of frames, and the motion also distracts the attention from the lack of definition. For preview purposes it is possible to use much greater degrees of compression than would be useful for finished images, and as you will appreciate this cuts down the disk storage space considerably, leaving room for more audio.

ISO

ISO (International Standards Organization) began during World War II when the United States and its allies had a requirement for standardization of interface characteristics. It is now a non-treaty agency of the United Nations and is a self regulating group with seventy-two members and eighteen non-voting developing countries. Member bodies may be private companies, voluntary organizations, and national or government agencies. ISO's remit covers agriculture, nuclear systems, fabrics, documents and more. Their intention is to be user orientated and to set standards that are 'for the good of the industry'. All ISO standards are reviewed every five years.

CCITT

CCITT (Comité Consultatif Internationale de Télégraphique et Télé-phonique) is now known as ITU-T, the Telecommunication Standardi-zation Sector of the International Telecommunications Union.

The CCITT was created from two predecessor organizations and was initially concerned with creating adequate volume on European telephone long lines. They issued recommendations to allow the compatibility of telecommunication services across national bound-aries, which are now often used as specification for the supply of telecommunications equipment.

MPEG2

Has anybody told you that there will be no sleight-of-hand magic shows on digital TV? Sleight of hand depends on people not seeing what the magician is really doing, despite it happening right in front of their eyes. This tells us something about the human visual system – we only see things that capture our attention, the rest is virtually ignored. Video data compression systems exploit this fact and only encode information that is likely to be useful, discarding as much as possible of the rest. Undoubtedly, therefore, any competent compression system will realize that viewers will only see the distraction created by the magician and not notice the actual manipulation of the cards, and therefore not bother to encode it, which rather defeats the object of the magic.

It is worth remembering why we need video data compression in the first place. Broadcast quality digital video recording generates a tremen-dous quantity of data – of the order of 200 Megabits per second (200 Megabits represents sufficient capacity for two and a half minutes of CD quality digital audio). Storing this on tape is difficult but possible; transmission of data in such quantities is and always will be prohibitively expensive in terms of bandwidth. Physical factors only allow a certain amount of bandwidth that is practical for terrestrial and satellite broadcasting. Therefore only a certain number of channels of a given bandwidth can be accommodated. We are faced with a trade-off between the number of channels and the bandwidth allotted to each one. In terrestrial analogue television, each channel requires 8 MHz of band-width, which is enough to accommodate a 5.5 MHz analogue video signal, mono or stereo analogue audio (depending on territory), digital stereo audio (again depending on territory), plus a safety margin. A 200 Mbit/s digital video signal would require a bandwidth of greater than 200 MHz, which plainly is far beyond the realms of practicality.

At this point, one might easily say why bother even trying to broadcast digital images if the data rate is so high? The answer is that it is possible with data compression to reduce the bandwidth, not just to the level of analogue video but much further – to around a sixth for images of comparable quality. The prospect of having six digital channels for every one analogue is very tempting. More channels means more programming, which means more production and more work for us all! There are of course other means of delivery. For example, a DVD Video disc is far too small to accommodate a sufficient quantity of uncompressed video data, therefore compression is absolutely necessary if a full-length feature film is to fit onto a single disc.

Syntax and semantics

Official MPEG documents frequently refer to syntax and semantics, which in language refer to sentence construction and the way in which words are used, respectively. MPEG2 is all about the way in which a number of techniques are applied in order to produce a data stream which a wide variety of equipment can understand. (MPEG, as you know, stands for Motion Picture Experts Group.) What MPEG2 does not specify is the equipment itself. Any design of encoder or decoder, as long as it can produce or work with an MPEG2 data stream, is satisfactory. This allows manufacturers the ability to research further, even after the standard is set in tablets of stone, and produce better equipment both for producers and consumers. This is in contrast with uncompressed digital video, where the image quality is defined in the standard and, aside from the A/D and D/A processes, can never get any better in any given format. MPEG2 deals with parameters such as image size, resolution, bit rate, and the nature of the MPEG2 data stream itself, without getting involved in the equipment necessary actually to make it work.

If there is an MPEG2 then obviously there must at some stage have been an MPEG1. Indeed there is, and an MPEG4 and MPEG7 too. (MPEG3, which dealt with high definition images, got lost along the way as it was found that it was more practical simply to extend MPEG2. MPEGs 5 and 6 seem to have disappeared also.) MPEG1 is an earlier and less ambitious standard upon which MPEG2 is based. The ultimate image quality and variety of applications of MPEG1 are limited – for instance it does not support an interlaced scan as used in television. Its use is generally restricted to applications where quality is not so much of an issue. MPEG2 is backwards compatible with MPEG1 so that existing MPEG1 software product does not become obsolete as MPEG2 prospers. Of course, an MPEG1 decoder would not be able to deal with an MPEG2 data stream.

Spatial and temporal compression

Spatial compression involves analysing a still image – one frame from a movie perhaps – and throwing away the parts that the eye cannot see. To examine this, first I need to mention the term 'spatial frequency'. Imagine a railing with vertical bars. The bars may have a spatial frequency of something like 10 bars/metre. Now think of a comb, which may have 5 teeth per centimetre giving it a spatial frequency of 500 teeth/metre. When you consider the eye's response to spatial frequency, then obviously you have to consider how far away the object is, so instead of measuring spatial frequency in cycles/metre, it is measured (for these purposes) in cycles/degree. Now you understand spatial frequency. It turns out that the eye is most sensitive at a spatial frequency of 5 cycles/degree, and hardly sensitive at all at 100 cycles/degree – the detail is too fine to be resolved. In the colour component of the image, the eye can only cope with detail as fine as 12 cycles/degree – anything finer and the gradations run into one another. One of the fundamental processes of JPEG (Joint Photographic Experts Group) still image compression, which is at the heart of MPEG2, is the discrete cosine transfer (DCT) where an image is converted into its component spatial frequencies. When this is done, those spatial frequencies that are irrelevant to the eye can be discarded. The next step is quantization where spatial frequencies are rounded to the nearest convenient value resulting in some loss of image quality because visible information has now been thrown away. Further steps pack the information digitally using techniques that require the fewest bits necessary to describe the image. The result is a still image which contains only a fraction of the data, but may look to the eye almost perfect, particularly if it is just one of a sequence of moving images.

Temporal compression involves reducing a sequence of images to key frames, and then only describing the differences between other frames and the key frames. Imagine for instance a soccer game just before kick-off. The first frame contains a lot of information, but as kick-off takes place very little changes until the camera moves or the director cuts to another angle. The grass, the stadium, the markings on the pitch are all in the same place and therefore do not need to be described again. In fact, even if the camera pans to follow the action, the data describing all of these things doesn't change, it just shifts position. From this we can see that there are many redundancies in a sequence of frames that can be removed completely while retaining virtually all of the real information content of the scene.

The MPEG2 data stream is divided into Groups of Pictures or GOPs. Each GOP consists of three types of picture. Intraframe (I) pictures are the basic information resource upon which other pictures will be based. All compression in an intraframe picture is done within the frame so that it can stand alone and be accessed randomly. Predictive (P) pictures are based on the previous I or P picture and contain only the differences between the two. Subsequent frames may contain information derived from predictive pictures. Bi-directional (B) pictures look forwards and back to the previous and subsequent P and B pictures and are encoded according to the data those pictures contain, bearing in mind that in all likelihood any bi-directional picture will be a sort of halfway house between the two pictures from which it derives much of its data. No other picture derives information from a bi-directional picture so any errors that accumulate are not allowed to propagate further. Each type of picture finds increasing opportunity to eliminate redundancies. The intraframe picture exploits redundancies within the frame itself; the predictive and bi-directional pictures find redundancies in the pictures on which they based, with the result that typically an I picture might contain 400K of data, a P picture 200K, and a B picture a mere 80K.

Look at the GOP as a whole (Figure 11.8). Each GOP must start and end with an I picture. In between are P and B pictures in an order set by the designer of the encoding system, not necessarily as in Figure 11.8. A GOP could consist of anything between eight and twenty-four pictures but twelve or sixteen would be more common. The frequency of I pictures

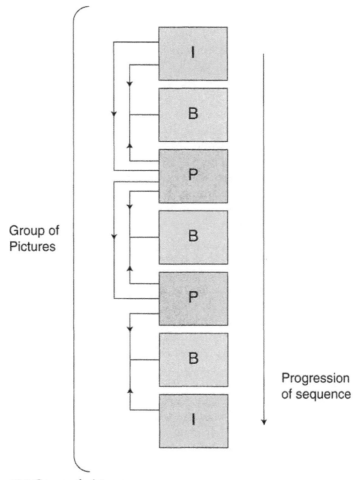

Figure 11.8 Group of pictures.

depends on the need to access the video stream randomly. For sequential viewing, such as a movie, random access is not as important as in a game where the action must start and stop according to the input from the player. Also, the nature and frequency of the transitions in the programme will determine the requirement for I pictures. Obviously a cut transition generates a large amount of completely new data, therefore an I picture is likely to be required since very little data, or none at all, can be carried over from the previous frame. Since B pictures depend on data from earlier and later pictures to be reconstructed, the order of the pictures in the data stream is not the same as they were shot, or will be displayed. It is better to send pictures upon which B pictures will be based, before their associated B pictures. Since this is a digital system and data can easily be buffered, there is no difficulty or problem with doing this.

Motion estimation prediction

It will not be long before our conversation is as peppered with the likes of 'macroblock' and 'motion estimation prediction' as it is now with 'decibels' and 'gain'. Each picture is divided into regions known as macroblocks where each macroblock is 16×16 pixels in size. Just think of it as a small part of the image because it is no more complex than that. If the camera pans, then the macroblock will change position on the screen, but if the scene remains the same, then the macroblock will be the same, and contain the same data. There is an obvious opportunity therefore to use this redundancy to reduce the burden on the data stream by looking for macroblocks which have simply moved, and describing where they have moved to in fewer bytes than it would take to describe the same macroblocks all over again. Motion estimation prediction is not used for I pictures, since an I picture is completely described with no reference to any other picture. But P and B pictures are both created using data derived from motion estimation prediction. For a P picture, the previous picture will be searched for matching macroblocks; for a B picture, both the previous and succeeding pictures will be searched. In the encoder, if a close match is found for a particular macroblock, then a motion vector is calculated which describes the offset, or where the macroblock has moved to. In addition a prediction error is calculated by subtracting each pixel in the macroblock from its counterpart in the previous frame (Figure 11.9).

All of this raises the question of where the encoder should search for matching macroblocks and how the search should be carried out, since obviously this could be a time-consuming process. Firstly a search area is defined, and a search carried out on a pixel or half-pixel basis. A half-pixel search means that adjacent pixels are interpolated, which results in a higher quality picture due to more accurate motion prediction, but

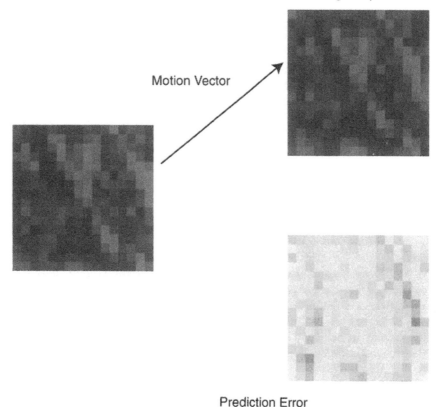

Motion Vector

Prediction Error

Figure 11.9 The prediction error is calculated by subtracting each pixel in the macroblock from its counterpart in the previous frame.

requires more computational effort. The manner in which the search is carried out may vary. The brute force way would be to compare each macroblock one by one within the search area (remembering that this process is carried out for every macroblock in the whole picture). Alternatively a telescopic motion estimation search can be done where every fourth macroblock is compared to look for one which is a close match, and when a close match is found its neighbours can be checked to see if they are any closer. A more common technique is to filter the image so that its area is effectively quartered before the search is carried out. This is known as a hierarchical motion estimation search.

Now you know what to look for in an MPEG encoded video stream. Where sections of the image remain largely the same for a period of time (a few frames for instance), they will tend to 'lock' because the encoder has decided that there is redundant data that can be discarded. Also, particularly where a scene has been shot using a hand-held camera, these 'locked' blocks (which may consist of a number of macroblocks) will tend to move as a whole. And when the data has changed sufficiently the blocks will unlock as new data is encoded only to lock again as the encoder seeks to reduce the amount of information so it can fit into the desired bandwidth. A good example of this can be seen in some news broadcasts where material is transmitted from around the world using fairly heavy data reduction. MPEG is of course capable of very much better quality than this when a higher bandwidth is available allowing more real rather than reconstructed data through to the viewer, but the parameters within which news broadcasters operate do not always allow for this.

Profiles and levels

No-one promised MPEG2 would be easy, but it certainly is versatile with seven Profiles and four Levels which can be combined in a variety of ways to suit the intended application. A profile is a subset of the full MPEG2 syntax and therefore defines how a data stream may be produced. A level defines basic parameters such as picture size, resolution and bit rate. A particular combination of profile and level would be written as Profile@Level. Several of the available profiles are hierarchical in that each is a subset of a more complex one above. The five hierarchical profiles, in increasing order of complexity, are:

> Simple
> Main
> SNR
> Spatial
> High

The levels, also in increasing order, are:

> Low
> Main
> High-1440
> High

Out of the twenty possible combinations, eleven are allowed, of which I shall describe a few of the more popular.

MainProfile@MainLevel (or **MP@ML**) is clearly intended to be the standard version of MPEG2. MP@ML supports interlaced video, random access to pictures, B pictures and employs a 4:2:0 YUV colour representation where the luminance component is given full bandwidth, the chrominance is subsampled both horizontally and vertically to give a quarter of the luminance resolution. Main Level has a resolution of 720 samples per line with 576 lines per frame and 30 frames per second and subjective tests have shown that MP@ML is equivalent to a conventional

PAL or NTSC image. **SimpleProfile@MainLevel** is similar but B pictures are not allowed, with the result that the decoder needs only a single frame memory and may be less complex, hence cheaper.

SNRProfile@MainLevel (or **SNR@ML**) is equivalent to MP@ML with the addition of a 'quality enhancement layer'. SNR Profile, which is said to be a scalable profile, consists of 2 bit streams, one of which may be decoded by an MP@ML decoder and the other providing better image quality through improved compression. For the technically minded, the enhancement layer uses finer quantization steps for the DCT coefficients, which basically means that each compressed frame is a better imitation of

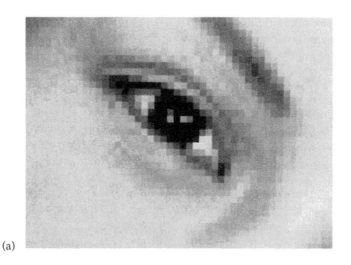

(a)

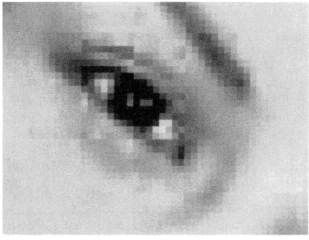

(b)

Figure 11.10 A highly magnified section of an uncompressed image compared with its compressed counterpart. The compression ratio in this instance is around 12:1. (a) Original; (b) compressed.

the original. It is a little like quantizing audio to 20 bits rather than 16, but then allowing the 16 bit signal to be replayed if for some reason the other 4 bits were garbled. The advantage of this in a transmission system is that if conditions are good with a good signal-to-noise ratio then all of the data may be received and a high quality image reconstructed. If conditions are poor then the base layer, which is more highly protected from transmission errors, may still be decoded to produce at least a good quality image. Spatial Profile is also scalable and in addition to the type of scalability outlined above different spatial resolutions are allowed at the receiver. The aim is to broadcast a high definition picture which is decodable by a standard definition receiver. **Spatial Profile** is currently only implemented at **High-1440 Level (Spatial@H-14)**. The lower layer has a resolution of 720 samples/line and 576 lines/frame, the spatial enhancement layer has a resolution of 1440 samples/line and 1152 lines/frame, and is also capable of 60 frames/second.

Although there is a **HighProfile@HighLevel** combination available which is capable of full 4:2:2 colour and even higher resolution than Spatial@H-14, it does not really lend itself to production applications since none of the hierarchical MPEG2 profiles were designed to withstand repeated decoding and re-encoding, which is obviously something that will happen often in post-production. Hence the **4:2:2Profile@MainLevel** was developed which has 4:2:2 colour, resolution equivalent to PAL and is robust enough to be repeatedly, to a reasonable extent, decoded and re-encoded. 4:2:2Profile@MainLevel is, as an example, used by Sony's Betacam SX system. Another non-hierarchical profile, **Multiview Profile**, is currently under development to allow two images from physically close cameras to be encoded to allow efficient transmission and storage of 3D moving images.

Applications

You may have heard fashion designers say that grey is the new black, or something similar. Well, MPEG2 is the new digital as far as video is concerned. MPEG2 will eat its way into every application of digital video, apart from perhaps the very top end of broadcast where transparent, meaning uncompressed, quality is considered essential. Cable and satellite TV broadcasters are obvious targets for MPEG2 encoder sales teams, since the whole *raison d'être* of digital television is that several digital channels can be squeezed into the bandwidth of one analogue channel, and at the same – or better – quality if desired. High Definition Television (HDTV) had been waiting in the wings for many years, lacking only the means of distribution at an affordable bandwidth. MPEG2 at last allows the means for high definition images to be seen in the home. DVD Video is a well known MPEG2 application which will allow variable bit rate coding. Variable bit rate coding increases coding efficiency and

Table 11.1 MPEG2 profiles.

Profile	
High	4:2:2 colour representation Supports features of all other profiles
Spatial	Spatial scalability SNR scalability Supports features of SNR, Main and Simple profiles
SNR	SNR scalability Supports features of Main and Simple profiles
Main	Non-scalable Supports B pictures Supports features of Simple profile
Simple	Non-scalable No B pictures 4:2:0 colour representation

Table 11.2 MPEG2 levels.

Level	*High*	*High 1440*	*Main*	*Low*
Samples per line	1920	1440	720	352
Lines per frame	1152	1152	576	288
Frames per second	60	60	30	30
Mbits per second	80	60	15	4

therefore the data may be stored more compactly. This will allow several different camera angles of the same scene, for instance, or different versions of the storyline so the viewer may interact. Video on Demand (VOD) is another application for MPEG2 where viewers in a hotel, and eventually the home, may select a programme of their choice rather than having to wait for it to turn up in the schedules. VOD calls for massive amounts of programming to be available which obviously must be compressed for efficient storage. MPEG2 is the answer.

CHAPTER 12

Digital television

Digital television in the UK

Did you know that roughly 80% of the information you receive through your television set is of absolutely no use to you? Yes, I know that sometimes it seems like 80% of the programmes in the schedule are repeats, cookery or gardening, but even if you are watching a particularly gripping episode of your favourite drama series, you can still only appreciate about a fifth of the information content that makes up the sound and picture. The human visual system is particularly selective in what it chooses to see. It has to do this in order to make a reasonable amount of sense of the world, and a vast amount of useless information is simply ignored. Existing video and television systems take advantage of this, for example by not recording or transmitting any fine detail in the colour component of the image, since the eye's sensitivity to detail in colour is roughly a third of its sensitivity to the fine detail in the brightness of a scene. This is due to the physical construction of the eye itself, but the brain also filters out irrelevant information. Once the brain has recognized the player with the ball, it is freed from the duty of facial recognition so you can give your full attention to marvelling at his playing skills.

Digital video

In digital video, the degree to which it is possible to manipulate the data means that information which the brain is not interested in can be discarded. For example, once you have seen the green grass of the football pitch, you don't need to be updated with how fast it is growing every twenty-fifth of a second. Similar information can be repeated over several frames, saving enormous amounts of data. Also, within a single frame showing the same football pitch, you don't need to see every blade of grass. Where there is not much difference between one part of the image and another, data can be averaged with hardly any perceived loss. In fact, when the savings due to all the possible means of discarding useless information are added up, one digital television channel can be squeezed down to around a fifth of the bandwidth of a conventional analogue channel, and the surprising thing is that it can look just as good and possibly better!

Since bandwidth is such a precious commodity (no-one is creating any more of it, except the cable companies), it seems almost a criminal waste to fill it all up with just four or five terrestrial analogue television channels. Why have five channels when we can have fifty? And with other methods of distribution it is not only possible, it is inevitable that in a few years time we will be able to choose from literally hundreds of channels. Some might say that they would rather have five good channels than a hundred channels of rubbish, but no-one complains about having too much choice in a bookshop. Television will become a bookshop where anyone who can find a willing publisher can have their say.

Broadcasting

DVB stands for Digital Video Broadcasting. Representatives of the major European broadcasters gathered together to form the DVB Project which defines the standards embodied in digital television.

Digital television can reach the home through four methods of transmission: terrestrial and satellite broadcasting, cable and – surprisingly – through the copper cabled telephone network. At the moment, it would only be possible to receive at best one VHS quality channel through your telephone line, but of course optical cabling and other technologies will eventually change this situation. The other three types of delivery, terrestrial, satellite and cable, are all very practical, and each has its own technical requirements. The common factor to all of these systems is MPEG2 coding, which can reduce the bit rate for a good quality picture to about 4 megabits/second. You may have read that MPEG2 takes longer than real-time to encode, which would of course make live broadcasting impossible. Fortunately, it is now practical to do MPEG2 coding in real-time so this is no longer a problem.

MPEG2 is not the only way the bandwidth requirement of a digital signal can be reduced. The way it is modulated for broadcast also makes a significant impact. Broadcasting digital information is not as simple as sending it down a short length of wire, as we do with our digital audio and video signals in the studio. You will certainly at some time in your TV viewing past have been troubled by ghost images caused by reflections from buildings, hills or even aircraft. These are annoying in analogue television, but potentially devastating for digital broadcasting.

To combat this, terrestrial digital television uses COFDM modulation, which stands for Coded Orthogonal Frequency Division Multiplex.

Briefly, this means that the data is spread, with error correction, over several thousand individual carrier frequencies (2000 in the UK). Just as the error correction of a digital recorder can compensate for a drop-out, so error correction in COFDM can build up a perfect picture from the data that survives the transmission/reception process and ignore any faulty data. The 'orthogonal' part of COFDM simply means that the frequencies of the carriers have been chosen so that they do not interfere with each other. Each carrier has a multi-level coding system where both the phase and the amplitude of the carrier are modulated. Compare this with AM broadcasting where only the amplitude of the carrier changes, and FM where only the frequency changes. It is not difficult to see that if you modulate both, then much more data can be contained in the signal.

The situation with satellite broadcasting is rather different. Since there is always a direct line of sight between the satellite and the dish antenna, there is never any multi-path problem. The difficulty is that the transmitter is 22 000 miles away (nearly three times the diameter of the Earth) so the signal is very weak by the time it gets here, only just enough to achieve a reasonable signal-to-noise ratio with very little safety margin. Modulating the amplitude of the signal is therefore very obviously out of the question. Only the phase is varied in QPSK (Quadrature Phase Shift Keying).

Once again this sounds a lot more complicated than it is. In QPSK, the phase of the carrier is allowed to swap between four states, each separated by 90 degrees. This is very similar in fact to NICAM digital audio in the UK where each state signifies two binary digits, thus increasing the data capacity. In fact, along with powerful error correction coding, this system has the potential of allowing smaller dishes of around 45 or 50 cm, which certainly makes them look less like a wart on the side of a house.

The fruits of all this clever technology are these: six channels of reasonably high quality television images can be contained within the bandwidth of one analogue terrestrial channel. If a lower picture quality can be tolerated, then of course more channels can be allowed. Alternatively, the full potential can be applied to one very high quality HDTV channel. Satellites have around twice the capability of terrestrial channels, and 12 standard digital channels or two HDTV channels can be accommodated within the bandwidth of one analogue channel. Cable systems have around the same capacity as satellite.

Possibilities

With such an immense improvement in programme bandwidth (for want of a better term) just around the corner, the big question is, 'What are we going to do with it all?' More of the same is the easy answer, but the obvious risk is that talent and resources will be diluted to the point that

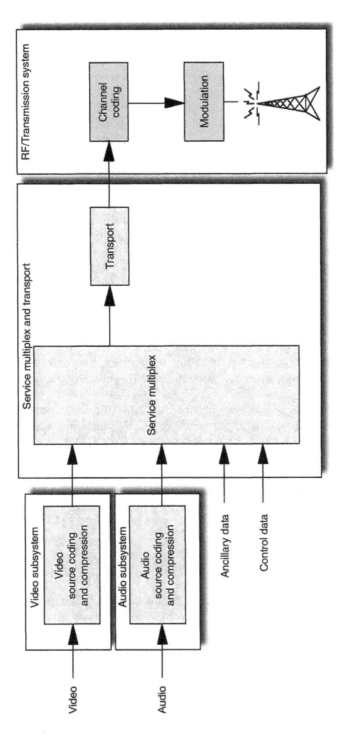

Figure 12.1 Digital terrestrial television broadcasting model. 'Service multiplex and transport' refers to the means of dividing the digital data stream into packets of information, identification of each packet or packet type, and multiplexing of all data types into a single data stream.

we have hundreds of channels of utter rubbish. It is not hard to think of suitable additional programmes that would complement existing output. A sporting event like the Olympic Games would be an obvious occasion where there are many events going on simultaneously, and minority sports enthusiasts normally have to make do with the shortest of visits to their preferred arena, usually after the event has taken place. Soccer enthusiasts will be very interested in additional analysis of the games, possibly including additional video material not included in the main transmission. Maybe one day fans will be offered two versions of a big game, with distinctly biased direction and commentary, as opposed to the even-handed approach broadcasters try to take now. Sport is obviously an area where there is a vast amount of material currently unbroadcast. Music is another, where a Glastonbury style music festival is currently shown only in fragments, when many fans would like to see the whole thing with a choice of any of the stages live, reruns of what they missed as it happened on the other stages, interviews with the artists, perhaps even archive material from previous years.

News coverage will also benefit from digital television. As we can hear on the radio and see on satellite now, a continuous news service is a very valuable option, allowing access to news when you want it rather than having to wait for it in the schedules. Regional programming too could be extended with more information, better tailored to each community. Education is another area where there is scope for vast amounts of programming, for which at the moment there is little time available.

And then there is always the archive. It's funny how it seems OK to buy a CD of music recorded 20 years ago, but a little bit shameful to admit to wanting to watch TV programmes of similar vintage that you remember enjoying. I wouldn't be at all surprised if eventually we might be able to watch an episode of Coronation Street (a soap with which about 20 million Brits are obsessed!) from 1965, or any other year since it began, and continue following it three times a week from that point on.

On demand

The ultimate state of digital television, in fact, will be Video on Demand, or near-Video on Demand more correctly. This is where the viewer is released from the constraints of the schedule and the same programme is broadcast at ten minute intervals on different channels so you can watch it when you like, rather than when the scheduler likes. This would not be practical for every programme, but certainly for major events it would be. Cable companies will be able to offer true video on demand where you order any programme to be delivered to your home exactly when you want it.

The consumer angle

Obviously, few potential viewers of digital television would be inclined to throw away their existing equipment and rush out and buy a new digital receiver and video. On the other hand, everyone who has satellite TV at home, or knows someone who has, is familiar with the concept of the 'set-top box'. The set-top box has proved to be perfectly acceptable to the public at large as a device to access extra services, whether cable or satellite channels, with smart card technology allowing viewing of encoded programmes, subscriptions, and pay-per-view.

To transform your existing television into a digital television receiver, all you need is a new set-top box. On the surface, this seems to be a very simple solution, but there are ramifications. It has been shown time and again that the public at large are very careful about what they will and what they will not accept from new technology, and any mistake at this stage could slow down the acceptance of digital television significantly. The first difficulty is that even though it is proven that people will accept a set-top box into their homes, is there any guarantee that they will accept more than one? Unless there is a good deal of co-operation between all sides of the TV industry, we might end up with a situation where you need one box for terrestrial digital television, another for satellite, and a third for cable, and multiples of these to cope with different service providers. There is only so much space on the set top. How high can we be expected to stack them? Hopefully, the set-top box will prove to be a transitional phase, and eventually it should be possible to buy TVs and videos with multi-standard digital decoders built in. I would speculate that when it becomes possible to buy a digital-capable portable TV for a couple of hundred pounds, the demise of the analogue channels will be inevitable.

I don't think that radio is going to be neglected when digital television comes to fruition. In fact, DAB (Digital Audio Broadcasting) for radio is with us now, using techniques similar to those employed in digital television. DAB will continue in parallel to digital television for people who choose to listen on portable sets or in their cars. Within digital television, whether terrestrial, satellite or cable, since digits are digits whether they describe pictures, sound or the balance of your bank account, the bandwidth used by one television channel could carry dozens of radio channels. The only potential problem is that the audio signal would eventually emerge from the set-top box, perhaps via the TV loudspeaker(s) or through a hook-up to the hi-fi. It sounds simple enough, but the public are used to the idea of radio coming out of radio sets and not through their TVs And how many people do you know – who are not sound engineers – who connect their TVs or videos up to their hi-fi? Not many, I would bet.

The coming of digital television could, in fact, signal the beginning of a new golden age of radio if people are prepared to give it a chance. The opportunities are there and hopefully people will be ready to benefit from them.

Widescreen

Quite apart from the potential multiplicity of set-top boxes, there is also a question of what the box should supply to existing receivers. Obviously, the introduction of a completely new system of broadcasting is a golden opportunity for widescreen television finally to gain mass acceptance. Digital television provides an opportunity for broadcasters to offer widescreen as an option in all of their programming, and it is the viewer who chooses the aspect ratio of the picture, regardless of the type of set they have. How this will be implemented is still open to debate. In an ideal world, additional coding would be transmitted so that if a viewer chooses to watch a widescreen movie in the old 4:3 aspect ratio, then the set-top box will pan to the most important part of the action, as is done now in telecine when preparing a film for 4:3 transmission.

Conclusion

Digital television is going to play a major part in our lives as producers and consumers of programming. There is every likelihood that we could have hundreds of channels of quality television and radio, widescreen, HDTV, video on demand and interactivity via cable or the Internet-bonanza for broadcasters, producers, facilities providers, advertisers, and couch potatoes everywhere!

Digital television in the USA

Just as digital television has taken its first stumbling steps towards flight in the UK, the USA has also started to move its television services over to digital delivery. And just as the government here promise that the analogue transmitters will be switched off sometime early in the next millennium, with safeguards for pensioners of course, the US government has a plan to do the same for terrestrial broadcasts on a roughly comparable timescale. We may take it as given then that we will all be watching digital television and good old analogue PAL and NTSC will be nothing but memories and material for those nostalgia programmes the BBC has always been so good at. In the USA, the progress of digital television (DTV) has been very closely linked to that of high definition television, or HDTV, under the general banner of advanced television (ATV) systems. Please excuse the acronyms but it is an acronym-strewn industry. HDTV has been around as an idea for many years. It is pretty obvious that current analogue television standards are not ideal (although many viewers seem perfectly happy) and there has long been a desire to move up to resolutions of around a thousand lines or more. Engineers naturally aspire towards higher technical standards, producers would probably prefer to see their creations in a more sparkling and crystal clear television medium, but the real impetus towards HDTV in the Western world has been political. Broadcasting is and always will be of prime interest to politicians since it is the most powerful medium of communication available. If anyone of a cynical disposition thought that the secret desire of politicians was to control our thoughts, then the politicians had better take control over broadcasting, which is historically what they have done in all the countries of the world. Once Japan had started, through the politically inspired doctrine of technical excellence, to take the lead in HDTV development in the 1980s, it was obvious in the United States that the spin-offs from all the research and development that would be necessary would place the Japanese ahead of the United States in all manner of communications technologies.

Also during the 1980s the US Federal Communications Commission (FCC) started to look at alternative uses for portions of the radio spectrum that had previously been unassigned. Their first thoughts were to allocate these frequencies to a category of use known as Land Mobile, which includes emergency services, delivery companies and others. This prompted broadcasters in the desire to stake a rival claim, not because they had any compelling reason to at that time, but simply because if a natural resource that is in limited supply is up for grabs, and bandwidth in the radio spectrum is the broadcaster's equivalent of a farmer's field, you might as well grab what you can. To justify their case, the broadcasters declared that they needed the bandwidth to develop HDTV services. The FCC listened and in 1987 appointed an advisory panel, the Advisory Committee on Advanced Television Service (ACATS), which was given the responsibility of examining the relevant technical issues and making a recommendation to the FCC on which system of ATV should be adopted. ACATS announced an open competition to develop an ATV standard, and among the twenty or so competitors was the Japanese analogue MUSE (Multiple Sub-Nyquist Sampling Encoding) standard which was already well developed. General Instrument, however, were able to demonstrate a DTV system, or at least the feasibility of such a system. It was evident to the FCC that DTV held great promise, although it was unlikely that that promise would be fulfilled in the short term. The FCC therefore delayed its decision until digital technology could be progressed towards a viable system.

In 1990, the FCC made some important decisions, firstly that whatever ATV standard achieved acceptance, it must be something significantly better than a mere enhancement of existing technology, and must be able to provide a genuinely high definition picture. Secondly, viewers should not, in the short term, be compelled to buy a new receiver and that conventional analogue broadcasts should continue alongside ATV transmissions. ACATS started a collaboration with a grouping of industry representatives, the Advanced Television Standards Committee (ATSC), and by 1993 had achieved a short list of four digital systems and one analogue. It was evident that the digital systems were significantly superior to the analogue, but they all had their shortcomings. Industry committees being what they are, it is not always possible to come to a decision simply with a show of hands, so a compromise was reached. Seven key players formed what came to be known as the Grand Alliance: AT&T (now known as Lucent Technologies), General Instrument, Massachusetts Institute of Technology, Philips, Thomson, the David Sarnoff Research Centre and Zenith Electronics. Inevitably, each had their own technologies to promote and were looking towards their own interests. It always was highly unlikely that there would be a single point of convergence, and the FCC is certainly not in the business of favouring any one manufacturer or developer. In fact the ATSC 'standard' as it eventually emerged, which is restricted to terrestrial broadcasting, is so wide-ranging in certain respects that it stretches the definition of the word beyond the vision of any lexicographer, as we shall see.

While the technical debate continued, Congress passed the Telecommunications Act of 1996 which paved the way for broadcasters to move forwards into the digital era. Existing broadcasters were granted a DTV licence and 6 MHz of bandwidth in addition to their normal 6 MHz analogue channel. This model is rather different to that used in the UK where a consortium (ONdigital) has been awarded the entire terrestrial franchise for the whole of the country. In the USA, broadly speaking, for each analogue channel there is now an additional 6 MHz of bandwidth which the broadcaster can use as they see fit, although it must be said that the rollout of DTV was always planned to be staged and it will be a while before full coverage is reached. One feature of the introduction of DTV in the United States which may seem unusual is that broadcasters are not allowed to wait and see whether it is going to take root – they have to get involved whether they like it or not. Digital terrestrial broadcasting in the USA officially started on 1st November 1998. Affiliates of the top four networks, ABC, CBS, Fox and NBC, in the top ten markets were required to be broadcasting digitally by 1st May 1999. Those in the top thirty markets had to be doing so by 1st November 1999. All other commercial stations have to be broadcasting digitally by 1st May 2002. To mitigate the element of compulsion, broadcasters are given a pretty free rein on what they can do with their 6 MHz, although the public interest standards of normal broadcasting still apply. If broadcasters choose to provide subscription services then they must pay a fee equivalent to 5% of gross revenues to Federal Government which is calculated to approximate to what they would have paid had the bandwidth been auctioned. Broadcasters are also required to ensure that the transition to DTV does not impinge upon the current notion that free over-the-air television should be available to all. At the end of the process, the analogue transmitters will be switched off on 31st December 2006. The Balanced Budget Act of 1997 does allow certain get-outs: analogue broadcasts will continue if any of the most significant broadcasters in a particular market

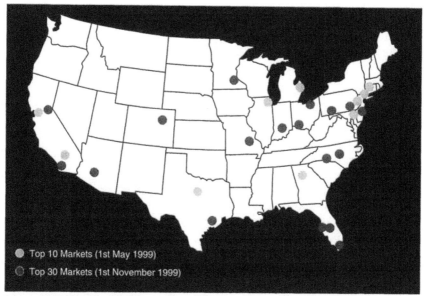

Top 10 Markets (1st May 1999)
Top 30 Markets (1st November 1999)

Figure 12.2 Digital television rollout. The top ten markets covering 30% of US households were scheduled to have been in place by 1st May 1999, the top thirty markets covering 50% of households by 1st November 1999. The remainder of commercial broadcasters have to be transmitting digitally by 1st May 2002.

has not started DTV broadcasts through reasons outside their control or if more than 15% of households in a market have for some unaccountable reason not made provision to receive DTV broadcasts (i.e. they still find the set they bought back in the 1980s perfectly satisfactory). One has to suspect that the second of these clauses will be the key and the FCC may have to go round the streets handing over free set-top boxes if people prove reluctant to accept the change.

As mentioned earlier, broadcasters are allocated 6 MHz of bandwidth to do with as they will, pretty much. You will remember that the original reasoning behind the transition to DTV was to enable high definition broadcasting. It seems, however, that there is a certain reluctance to head in this direction and most broadcasters find the alternative of multi-casting much more appealing. Since a DTV broadcast occupies very much less bandwidth than an analogue broadcast of broadly equivalent quality, it is possible to broadcast four or five DTV streams within that 6 MHz channel. More channels of course means more opportunity to sell advertising. Simple market economics apply. If you are broadcasting one HDTV channel, and your rival is broadcasting four SDTV (standard definition television) channels, who will earn the most revenue? The choice between SDTV multicasting and a single HDTV channel is not fixed. It would be perfectly possible to multicast during the day and switch over to HDTV during primetime. How the viewers will react to this kind of variability is a good question, as would be their interest in having a certain portion of the 6 MHz bandwidth given over to data services.

Technical issues

Among the FCC proposals in 1996 were the technical standards devised by the Grand Alliance. These include standards for scanning, video compression, audio compression, data transport and transmission. The proposals on scanning are the most diverse with not one standard but eighteen! Currently in PAL television we have 625 lines per frame, transmitted at 25 frames per second. In NTSC there are 525 lines per frame transmitted at 30 frames per second. Simple and easily understood. For DTV there is a hierarchy of scanning standards which starts at resolutions of 480×640 and 480×704 pixels which are comparable with good old NTSC. In the middle there is a standard of rather higher resolution with images consisting of 720×1280 pixels. Right at the top sits the true HDTV resolution of 1080×1920 pixels, a single channel of which compresses nicely into the 6 MHz available bandwidth. But there is more to scanning than resolution. Some of the standards call for interlaced scanning, as found in analogue television, where half the lines of the frame are displayed as one 'field', then the other half are slotted in between, making up the full frame. Interlaced scan was adopted in the early history of television as a means of reducing flicker without taking up excessive bandwidth. Modern computer displays do not need to interlace the lines and employ what is known as progressive scan, where the lines are displayed sequentially from top to bottom. Both interlaced and progressive scans are allowed under the ATSC standards and it presumably remains for the receiver to sort out which is which. Another variation is the aspect ratio, and both the old 4:3 ratio and the newer 16:9 are allowed. Frame rates may be 30 fps or 60 fps, although 29.97 fps is expected to remain in use in production.

Video compression is the familiar MPEG2, and audio is in the respected Dolby Digital format allowing 5.1 channels. Transmission of DTV signals in the USA employs a different system to the COFDM (Coded Orthogonal Frequency Division Multiplex) used in Europe. COFDM was considered, but the 8-VSB system was found to be better suited to the needs of the American market. In fact, the two systems are involved in a running battle for emerging digital television markets all over the world, but that's another story entirely. Whichever system of broadcast is used, the problems remain the same. Terrestrial broadcast systems are subject to interference from other radio sources, and from multi-path reception resulting in a 'ghost' signal. Interference and multi-path artefacts are visually irritating in analogue systems, but they have the potential to destroy a digital signal completely if not held in check. In US-style digital television, the MPEG2 encoder produces the so-called transport stream layer which runs at a data rate of 19.39 Mbits/s which, as previously mentioned, must be slotted into a 6 MHz bandwidth. Turning the transport stream layer into a digital TV signal is the work of a device known as the 8-VSB exciter. Figure 12.1 shows a block diagram – let me pick out a few of the more interesting stages in the process. After decades of research and experience in broadcasting, engineers have determined that it is not efficient to allow the signal to aggregate into 'clumps' of frequencies within the signal's bandwidth. This is inefficient since it means that while parts of the spectrum are fully utilized, other parts are not. Thus, to achieve optimum efficiency the signal must be spread out very evenly over the entire allotted band and become almost noise-like in its energy distribution. Obviously a TV signal starts off with a very regular structure so the function of the data randomizer is to take this regular structure and even out the peaks and troughs in the distribution of energy. This also helps prevent DTV signals from interfering with old-fashioned NTSC receivers. Following this, the Reed-Solomon encoder provides error correction according to a well established system in common use since the introduction of compact disc. To the 187 data bytes of each MPEG2 packet, an additional 20 parity bytes are added which allow the correction of errors of up to 10 bytes in each packet. If the error rate is higher than this then the packet has to be discarded, which does not necessarily lead to an obviously visible artefact since the MPEG2 decoder in the receiver or set-top box has additional means of disguising errors that do make it all the way through. The data interleaver works in exactly the same way as the interleave process in digital tape systems. In digital tape, it is appreciated that any flaw or drop-out in the tape could, left unchecked, destroy a considerable amount of data, including error-checking codes. To avoid this, data is held in a buffer memory and spread out over a wider area of the tape. Thus any dropout on the tape is not allowed to destroy all of a particular section of data, but because of the interleaving process causes minor errors over a wider spread of data in the hope that these errors can be completely corrected. In the 8-VSB exciter, the interleave process spreads the data out in time rather than over a wider area of tape, to the tune of some 4.5 ms, so that any burst of interference will, in the end, have minimal effect on a viewer's enjoyment.

Getting a little bit more complicated now we come to the Trellis encoder. This is another form of error correction but it works in a distinctly different way to Reed-Solomon coding. Reed-Solomon coding works on blocks of data and each block is treated as an entity within itself. Trellis coding on the other hand tracks the progress of the data stream and assesses the accuracy of the data at any one moment by comparing it with past data, and also future data (it is able to do this since the current data is buffered – there is no crystal ball, not even in digital TV). Trellis

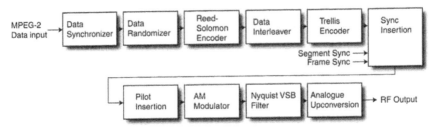

Figure 12.3 8-VSB exciter.

coding has, quite usefully, been compared to following footprints in sand. You could follow one clear set of footprints easily, but what if the path is confused by other footprints (interference)? Only by looking ahead and checking backwards will you know which is the correct path to follow. The other paths, which represent the interference, can be ignored. The trellis coder splits up each incoming data byte into four 2-bit words. From these 2-bit words it creates a 3-bit word describing the transition from one 2-bit word to the next. Three bits are capable of expressing 8 different levels of information, and this is where the '8' of 8-VSB comes from. A 16-VSB system was also developed, and presumably other variations are possible. The output of the trellis encoder consists of 828 eight-level symbols for each of the original 187 byte MPEG2 data packets. Following the trellis coder the signal is manipulated to make it easier for the receiver to pick up and lock onto. Sync pulses are added to identify each 828 symbol segment, and also the data streaming is organized into data frames consisting of 313 segments which on reception form the input to the receiver's complementary trellis decoder, error correction system and MPEG2 decoder. The level and width of the sync signals is very carefully calculated that they do not take too much bandwidth away from the all-important data. Even so, sync is recoverable down to an RF signal-to-noise ratio of 0 dB, which means that the noise is as strong as the signal itself. Data is only recoverable down to 15 dB but this margin helps the receiver lock onto the signal during channel change or momentary signal drop-out. In addition to the sync signals, one further manipulation of the signal is a d.c. offset which maintains a data-independent component which allows the receiver to lock on more easily.

The baseband signal incorporating eight discrete levels, d.c. offset and sync pulses is amplitude modulated (AM) onto an intermediate frequency (IF) carrier. You will recall from modulation theory that in amplitude modulation two sidebands are developed on either side of the carrier frequency, each being a mirror image of the other. This of course is wasteful of bandwidth since effectively the same information is being transmitted twice. In order to correct this, the lower sideband is almost completely suppressed leaving just a vestige of its former self behind. Vestigial sideband transmission is already used in NTSC broadcasting and is in itself nothing new, although its application to digital broadcasting proves that old techniques still have value. At the end of this process, the IF frequency is up-converted to the assigned channel frequency and broadcast, hopefully to millions of eager viewers.

As in UK digital television, doubts have certainly been expressed over whether the public really want it or need it, but the fact is that DTV is there in reality and although I predict that the ultimate analogue switch-off will be delayed, sooner or later it will go and analogue television as we know it, after some sixty or so good years of information, education, entertainment (and commercials!), will finally come to an end. Gone but not forgotten?

CHAPTER 13

Film

In the audio industry we are used to the idea of the new replacing the old. A new technique or piece of equipment is developed and gradually we all, or nearly all, move over to using it. We do this because the new always has an advantage over the old and – of course – provides within its universal set of capabilities a complete subset of what the old equipment could do, leaving out nothing at all. Well, we might wish that were the case because always when a new audio technique or piece of kit comes along, and we adopt it with enthusiasm, we find ourselves occasionally pining for the old days and those few things we cannot now achieve. Although the glossy advertisements in magazines portray a state of continual change for the better, as they would and as it mostly is, the true seeker for perfection will select whatever technique is appropriate for the job, whether it was invented yesterday, or twenty years ago.

I say all this because sometimes I feel that we are told so often that new is better than old that we end up not daring to disbelieve it ever. Yet once in a while a particular area of technology may be so close to perfection that it will be a long time before something more modern can adequately replace it. Like valve microphones for instance: only in the last few years have manufacturers revisited valve technology and have begun to accept the fact that for certain applications, transistorized mics just don't deliver, even though they may be smaller, lighter, more reliable and less expensive to manufacture. This is true of film also. We may assume that since film is such an ancient medium then it must be long past the end of its serviceable life and that electronic images will soon take its place. This assumption would be very remote from the truth of the matter, which is that film is alive, well and very active. Unless we are prepared to put up with less than film can offer, I would expect it to be around for decades to come, and in this chapter I shall explain why.

A brief history of film

Just as there is no one person who can truly be said to be the sole inventor of television, there is no-one who can be solely credited with the invention of film, although if Thomas Alva Edison were alive today he would surely be claiming to have done so. You already know about the eye's persistence of vision, where an image that falls upon the retina will be retained for a short period of time. This makes it possible for the brain to fuse a sequence of still images into smoothly flowing movement. In the period from about 1835 there were a number of devices based on this principle where a series of drawings of a simple motion could be repeatedly viewed in sequence through a contraption using slots or mirrors. These devices were given exotic names such as Zoötrope, Praxinoscope, Phenakistiscope, and one name which we still use today for a different piece of apparatus – Stroboscope. Of course, no-one could then imagine adapting these for anything like film because photography was still in its very early stages. It was not until the 1870s that a photograph could be exposed in a fraction of a second and moving photographic (as opposed to drawn or painted) pictures became a possibility. Around this time there was a fevered argument among the intelligentsia of the Western world (and everywhere else in all probability) over whether a galloping horse always had at least one foot on the ground, or whether at certain times all four feet were in the air. If you imagine a world without slow motion photography then you will realize how difficult it would be to determine this for sure. In 1878 San Francisco based Englishman Eadweard Muybridge devised a method to prove the argument one way or the other (to settle a bet of $25 000). He set up a line of twenty-four cameras whose shutters would fire in sequence, triggered by trip wires, and capture the motion of the galloping horse for analysis. (I don't suppose it is of much technological relevance to us, but the horse's feet were found at one point all to be in the air. Of course the horse knew this already.) Around this time other workers were inventing methods of recording sequences of pictures, one of which was called the photographic gun and recorded twelve images in one second on a circular plate. In France in 1887 the idea was conceived of recording the images on a strip of photographic paper, so all the elements of motion pictures were now in place.

Although the race towards moving pictures was well and truly on in all the industrialized countries, the work of Edison is very significant – although some would say that his ego, and eagerness for litigation, place his perceived reputation higher than it deserves, at least as far as film is concerned. His first thoughts on moving pictures were to present visual images to accompany the sounds produced by his phonograph, which he was already selling successfully. He first experimented with small pictures arranged in a spiral path mounted on a cylinder, the same arrangement as the phonograph, but in 1889 one of his assistants, William K.L. Dickson, according to one version of a much repeated story, attended a demonstration of George Eastman's early snapshot camera ('You press the button, we do the rest.') which used a flexible strip of film some one and three-eighth inches wide.

Flexible film was in fact the missing link, and Dickson recognized this. Dickson ordered a camera so that he could take a closer look at the film, and after successful tests Dickson visited Eastman in Rochester where they discussed manufacturing a wider film specially for Edison's motion picture project. Edison, however, kept the purse strings pulled tight and opted for the 35 mm gauge because it was cheaper. Edison subsequently patented the Kinetograph camera and Kinetoscope viewer in 1891. The camera was capable of shooting six hundred images on a flexible film 50 feet long at a frame rate of around ten per second. Surprisingly, Edison did not go on to develop a projector and remained content with his single-user Kinetoscope viewer. We may see this as an error of judgement in hindsight, but the Phonograph was also a single-user device with an earpiece rather than a horn and perhaps Edison thought it appropriate that the Kinetoscope should be deployed in the same way.

Meanwhile in France . . .

Some would say that in the field of jazz music, France is second only to the United States in originality and influence. You could say the same about cinema, except that the French language does not travel as widely as English, which does impose a certain restriction, and the style of French films is another world compared with mainstream Hollywood products. According to some estimates, France has the greatest number of cinema screens of any European country, and this may be due in part to the country's significant involvement in the early development of motion pictures. First and foremost, the words 'cinema' and 'cinematography' derive directly from Léon Bouly's Cinématographe camera/ viewer of 1892. These words fortunately trip off the tongue just a little more easily than the many possible alternatives, which include the rather more difficult Electrotachyscope and Eidoloscope – perhaps some inventions are just destined to fail. Central to France's efforts in the development of motion pictures were the Lumière brothers, Auguste and Louis. They studied Edison's work closely and developed his Kinetoscope viewer into a true projector and named it, unoriginally, the Cinématographe (poor old Bouly had allowed his patent to lapse). The Cinématographe, and other similar devices, demonstrated in the last half-decade of the nineteenth century that there were profits to be made from showing moving pictures to a paying audience. In 1895, the Lumières presented a sequence of twelve short movies lasting in total around twenty minutes to a small Parisian audience. Apparently the most popular was of a train pulling into a station! (In another presentation in the USA in the same year, history records that during the showing of a film of waves washing onto a beach, people from the

Figure 13.1 The Lumière brothers' single perforation per frame 35 mm format.

front rows of the auditorium jumped out of their seats to avoid getting wet!) Later showings packed the house.

The Lumière brothers, following the lead of Edison, also used film one and three-eighth inches wide, but in deference to French tradition called it 35 mm. A standard was born! 35 mm film was subsequently ratified in 1916 as an official standard by the forerunner of SMPTE, the SMPE, at their first meeting. (Guess why there's no 'T'!) 35 mm film was actually already something of a *de facto* standard so SMPE were not really doing anything more than rubber stamping it. Also standard were the frame dimensions of 24 mm × 18 mm and the frame rate of approximately 16 fps. Apparently audiences were so amazed at seeing anything like moving pictures that they were not at all bothered by the jerkiness of the images. A three-bladed shutter in the projectors of the time fortunately raised the flicker rate to 48 Hz, which is still the standard for film. The Lumière brothers' single circular perforation per frame had to give way to Edison's four sprocket holes per frame and over a short period this effectively became a worldwide standard. In fact the width of motion picture film has been quoted as being the only universal standard of measurement that is recognized in every country of the world, and this may well be true. What is certainly true even now is that you can shoot a 35 mm film and project it anywhere in the world. Try that with video!

The coming of sound

Early sound-for-film techniques often attempted to synchronize two obviously incompatible media (some things don't change!). The most elegant solution would have been to record the sound on the film itself, which doesn't guarantee no loss of sync ever, but there are definitely fewer things to go wrong. Sound on film was invented as a concept as early as 1888 by Eugene Augustin Lauste, a Frenchman who worked with Edison but later transferred his operations to England, but the electronic technology was not by then available to make it work. Lauste's idea was to project light through two gratings of clear and opaque bars. One grating would be stationary while the position of the other would be modulated by the audio signal, thus varying the intensity of the light that would fall on the film. On projection, a light would be shone through the resulting sound track and the varying degrees of light and shade would be picked up by a selenium photosensitive cell. The principal difficulty was that the system was too heavy to be successfully modulated by the meagre output of a microphone of the period and several years passed without any substantive achievement. Nevertheless, Lauste persevered and eventually devised a workable synchronized system where half of the film area was devoted to the soundtrack. Some might say that Lauste got his priorities right, and his work was an important foundation for the optical sound system still in use today.

Despite Lauste's achievements, the first practical synchronized sound system was the Western Electric Vitaphone system. In this system, sound was recorded onto an old-style shellac disc (vinyl was still decades away) and replayed in approximate sync with the projector. Approximate is a key word here since initially the quality of synchronization was entirely up to the skill of the projectionist. Soon, a fully mechanical arrangement was introduced where the film and disc were started at the correct point and, as long as the stylus didn't skip, lip-sync could be maintained for the whole of the reel. The first feature film produced with Vitaphone sound is commonly recognized as being The Jazz Singer. Although it was by no means the first Vitaphone film – earlier productions including Don Juan had music and sound effects – The Jazz Singer contained dialogue sections which indubitably confirmed its status as the first 'talkie', and the film world was never the same again. In particular, the standard film speed of 16 frames per second, which in fact was often varied at the whim of the director or camera operator, was reset to a precise 24 frames per second. Naturally, the speed of sound playback must be constant, and identical to the speed of the recording, so we can thank Stanley Watkins of Western Electric for this.

Significant though the Vitaphone system obviously was, it could not have continued indefinitely. Having sound and picture on separate media is inevitably a recipe for disaster as everything depends on the projectionist getting both sets of reels and discs in the right order, and of

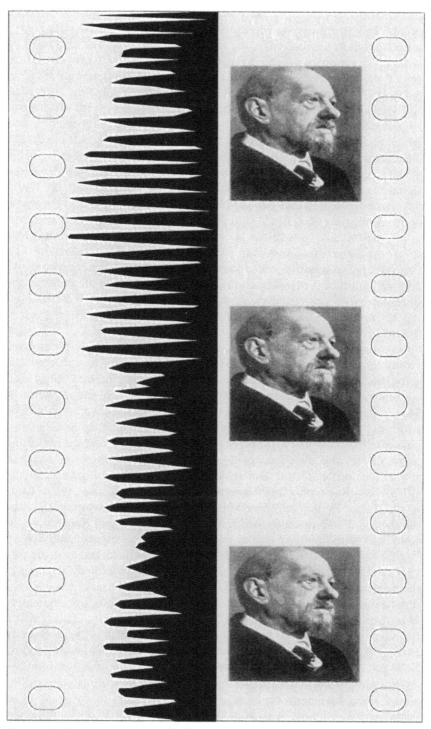

Figure 13.2 Eugene Lauste's optical soundtrack.

course a record can easily skip. Breakages were also a far too common event. The natural answer was to rekindle enthusiasm for Lauste's idea of sound on film. Concurrent with the development of disc recording, a number of companies and individuals were hard at work on optical sound. In particular, Theodore Case of General Electric had by 1920 developed a photographic recorder for radio telegraphy. By 1928 RCA and Western Electric both had optical sound on film systems working to a reasonable standard. Changeover to optical sound was progressive and by the mid-1930s most studios and theatres were optically equipped. One interesting feature of optical sound is the difference between variable density recording and variable area. In variable density recording the signal is represented by various shades of grey, while in variable width the sound track is always completely clear or completely black, the width of the clear area giving the instantaneous level. The film stock of the time, as you will realize, was rather poor at representing greys accurately and so the variable density system was prone to extreme distortion. The variable area (sometimes called variable width) gave a cleaner sound and became standard. Curiously enough, it didn't matter at all as far as the projector was concerned because the only thing its photocell cared about was the amount of light that shone onto it through the soundtrack area of the film.

One major problem of optical sound was noise. Film is naturally a grainy medium, and this random granularity leads to noise in the soundtrack, and in the picture too you will have noticed. Dust and scratches make the problem even worse. To combat this, the Academy of Motion Picture Arts and Sciences proposed that movie theatres should be equipped with filters to remove high frequencies, where the noise is most objectionable. 'High frequencies', they determined, meant anything above 6 kHz, and the effects of the filter were noticeable down as far as 1 kHz. This unsatisfactory state of affairs continued until the mid to late 1970s when Ray Dolby and his remarkable noise reduction system came to the rescue. (If you saw Star Wars in a good theatre when it was first released you will certainly still remember how amazing the sound was compared with anything else around at the time. I certainly still do!)

With the introduction of sound, since the 35 mm format was already very well established, no-one ever suggested changing the dimensions of the film to accommodate the addition of a sound track. Instead, the dimensions of the image were reduced from 24 × 18 mm to approximately 21 × 16 mm, which was known as the reduced aperture to distinguish it from the original full aperture. Since the frame rate is now 24 fps, the film runs at eighteen inches per second, and by simple calculation you can find that a standard 1000 foot reel will last for just over ten minutes. Another important standard is the separation between the picture and the sound along the length of the film. It will be evident that it is not at all convenient to attempt to pick up the sound in the gate of the projector. The sound advance, as it is known, is standardized at twenty frames the world over. This separation allows for stop-start motion in the gate to project steady rather than blurred images, and the smooth motion that is obviously essential for sound. Amazingly enough, optical sound is still well and truly with us and virtually any film you go to see in your local movie theatre or out-of-town multiplex will have an optical sound track. Sounds pretty good still doesn't it? (Apart from the occasional pops and crackles.) At much greater expense, it has been possible since the 1950s to have magnetic sound tracks on the print shown in the theatre. This can be done when very high quality 70 mm presentation is required, but I doubt if it will survive as the new digital cinema sound techniques battle it out to be the new standard. Magnetic sound as part of the production process is still alive and just about kicking and will probably continue in some shape or form somewhere in the world until the equipment turns to rust. Magnetic sound had its day, particularly in relation to the widescreen movies of the 1950s and 1960s, but the cost of adding a magnetic stripe to each and every print and then copying the soundtrack onto it always was prohibitive for general release. Most movies released in widescreen format with magnetic soundtracks were subsequently put on general release in 35 mm optical sound versions, so most people never experienced them at anything like their best.

Figure 13.3 Modern stereo optical soundtrack.

Widescreen

In the early days of motion pictures, the aspect ratio of 35 mm film was originally set at 1.33:1, or 4:3 in round numbers. You will recognize this as the aspect ratio we see on standard television today. The need for a wider screen image to match the field of view of the human visual system has long been recognized and pursued in both film and TV. In film, there are five ways to produce a wider image than 1.33:1:

- Use a wider film. This is likely to be incompatible with the projection equipment in general use, although it has to be said that the adaptability of motion picture projectors has proved impressive over the years.
- Reduce the height of the image on the film and use a wider angle lens on projection. This will reduce the picture quality as a smaller area of film is being asked to cover a larger screen.
- Squeeze the image horizontally onto the film and then stretch it out on projection. This reduces the image quality in the horizontal direction slightly, but maintains vertical resolution (and in fact the loss of horizontal resolution may be compensated by a better use of the available film area).
- Run the film sideways to allow a wider image.
- Use multiple projectors. If one projector does not give a wide enough image – simply use more of them!

All of these methods have been used in the past and only one is not in current use today for theatrical presentation (apart from movie museums of course).

Widescreen using larger than normal film has a surprisingly ancient history, dating back all the way to the late 1920s and early 1930s with the Fox Grandeur system using a 48 × 22.5 mm frame on film 70 mm wide (leaving enough room for a 10 mm soundtrack!). The Fearless Superfilm camera, which had to wait more than twenty years for success, was designed for 65 mm stock but could be adapted to use 35 mm film, or to special order to virtually any of the other wider gauges that were in development at the time. Unaccountably, the early introduction of widescreen equipment did not lead to its general adoption. Perhaps the public needed a demonstration that would truly amaze them, rather than mere incremental improvements.

That amazing demonstration was provided by Cinerama, the brainchild of Fred Waller who developed the Waller Gunnery Trainer that used five cameras and projectors and was used during World War II to give trainee anti-aircraft gunners a target covering a realistically wide field of view. (We tend to think of simulators as being a recent development but obviously this is not the case.) A system similar to this had already been demonstrated at the 1939 New York World Fair. After World War II, the system was refined under the name Vitarama to three cameras, three projectors and a characteristic curved screen. The aspect ratio could approach that of three standard image widths added together, minus an allowance for a degree of overlap. With the assistance of sound specialist Hazard Reeves, the Cinerama system was born with five channels of audio coming from behind the screen and a further two in the auditorium. And we think that 5.1 is modern! The first Cinerama movie, This is Cinerama, was shown in 1952 to amazed audiences – it would have been a rather more impressive spectacle than the television images of the time I suspect. Cinerama did indeed have a degree of success, but the problem was that theatres required substantial conversion to accommodate the enormous curved screen. Also, the 146 degree angle of view of the camera meant that direction was difficult, as was lighting. In the end only seven movies were made in the format in the ten years from 1952 to 1962: This is Cinerama (1952), Cinerama Holiday (1955), Seven Wonders of the World (1956), Search for Paradise (1957), South Seas Adventure (1958), How the West Was Won (1962) and The Wonderful World of the Brothers Grimm (1962). When Cinerama Inc. was eventually bought by Pacific Theatres, the three-strip process was abandoned with the result that a number of movies that claim to be in Cinerama actually are not true Cinerama but other widescreen formats. These include It's a Mad, Mad, Mad, Mad World (1963), 2001: A Space Odyssey (1968), The Greatest Story Ever Told (1965) and Khartoum (1966).

Although Cinerama was apparently very well received, it was impractical as it could only be presented in specially adapted cinemas. Another option was pursued, as early as 1928 by Henri Chrétien of France, where the picture is shot with a lens which has a different magnification in the vertical and horizontal directions. This technique is known as anamorphosis (see Figure 13.5) and such a lens is called an anamorphic lens where a conventional lens would be a spherical lens. The resulting film is projected using a similar lens and the distortion is corrected. You will undoubtedly recognize the name CinemaScope, which used anamorphosis to achieve an aspect ratio of 2.66:1 – twice as wide as a standard 35 mm image. Although 35 mm film was retained, the width of the sprocket holes was reduced to make room for four magnetic sound tracks (quadrophony in 1953!). The frame height and four sprocket holes per frame standards were retained so the film could be projected without too much modification of the equipment. Although CinemaScope was judged to be a technical success, separate optical sound prints had to be made for cinemas that were not prepared to install the necessary magnetic sound equipment, and it proved that directors were not always willing to compose their shots to suit the extremely wide format. It became apparent that some compromise between the 1.33:1 and 2.66:1 ratios was necessary. (Film experts will undoubtedly pick me up on the 2.66:1 ratio I have quoted for CinemaScope: there was another ratio, 2.35:1, which was used for prints with both optical and magnetic soundtracks. The 2.35:1 ratio is also used in the Panavision format.)

Anamorphosis is a clever optical trick that certainly works, but there is a simpler way to achieve widescreen: simply chop off the top and bottom of the image! It sounds too simple to be true, but it isn't. Many movies are made in the so-called Super 35 format with an aspect ratio of typically 1.85:1. A 4:3 image is shot in the camera with foreknowledge that the upper and lower portions are not going to be seen in the cinema. The centre strip is anamorphically printed for exhibition. Of course a certain area of the camera film is wasted and a smaller negative image is asked to cover a large screen area, but it is thought by some directors that if sufficient care is given to all technical aspects of the production process, then the results are actually superior to anamorphic camera formats. Let's face it, you didn't quibble about the technical quality of James Cameron's Titanic (1997), did you? One bonus feature of Super 35 is that the additional image areas at the top and bottom are available when the movie is shown on TV or transferred to video or DVD in the 4:3 aspect ratio. With care, these can be used to avoid losing important action at the edges of the frame. It is also possible to shoot a format similar to Super 35 with three sprocket holes per frame rather than the normal four. Cost-conscious accountants appreciate the saving in stock, as do those who care for the environment.

Sideways look

To obtain a better quality wider picture the most obvious method is to increase the size of the original camera image. Apart from increasing the overall size of the film, this can also be achieved by running the film sideways, as in the VistaVision system which used a spherical lens and 35 mm film. VistaVision was used on a number of big name movies including White Christmas (1954), High Society (1956) and Vertigo (1958). VistaVision's great advantage was that although special cameras were used for shooting, the image was printed down to 35 mm for presentation, meaning that it could be shown in any theatre.

The best quality currently available (occasionally) in mainstream film is 70 mm, which simply uses the brute force technique of increasing the size of everything. A number of formats of around this size were developed, one of the most significant being Todd-AO. Todd-AO was the brainchild of Mike Todd, Broadway producer and a partner in the Cinerama venture. His dream was to produce a Cinerama-like image using a single strip of film, and devised a system which employed 65 mm film in the camera running at 30 frames per second to minimize flicker. A spherical lens was used. The 65 mm camera negative was subsequently transferred to 70 mm print stock to allow room for four magnetic sound tracks. Of course, the negative could be printed down to 35 mm with optical sound as required for smaller theatres. Oklahoma! (1955) was the format's first success, and a massive success it was too. One of the drawbacks of shooting at 30 fps was that the movie had to be shot twice – once in Todd-AO and again in CinemaScope. A rather inelegant process. Other significant Todd-AO productions include Around the World in 80 Days (1956) and South Pacific (1958). By this time, Mike Todd had left the company and South Pacific was shot at the conventional frame rate of 24 fps. Around the same time Metro-Goldwyn-Mayer were working on a system known as Camera 65 or Ultra Panavision. Ultra Panavision uses 65 mm camera negative stock once again, but this time with anamorphic lenses to produce a wider image. Raintree County (1957) was the first Ultra Panavision movie, followed by Ben-Hur (1959). Super Panavision 70 was the spherical-lensed version and is now pretty much the standard in 70 mm production, though 70 mm production is rare these days. Exodus (1960) was an early Super Panavision 70 production. Far and Away (1992) and Hamlet (1997) also used this format.

If 70 mm production is now rare for feature films, for special exhibitions there is another format, the achievement of which is on an altogether grander scale – Imax. Imax uses 65 mm running sideways in a similar manner to VistaVision to give an image more than twice the area of a conventional 65 mm frame. This can be subjected to the anamorphic treatment too, to become Imax Dome, or Omnimax, where the audience sits under a hemispherical screen and is bombarded all round by the action. Further extensions to the technology include 3D and Imax HD where the frame rate is an incredible – for such a large format – 48 fps!

Colour

Considering that movies were still often being made in black and white as late as the 1960s, it is amazing how early colour moving pictures were possible. And films such as Gone With the Wind (1939), and The Wizard of Oz (1939 also, although only some of it is in colour) have a particular 'look' that has stood the test of time well. As you know, to produce an illusion of full colour requires a combination of three images that will stimulate the red, green and blue sensors of the retina. Early systems used 'additive colour', either by projecting sequential red, green and blue images, or by combining the three images optically, like some video projectors do. Subtractive colour is a better option because three images in complementary colours (cyan, magenta and yellow) can be combined in a single frame. The Technicolor company was active in the field since 1917 and the term has become a synonym for all things gloriously coloured. Their first system was additive but their second, from 1922, used a two-colour subtractive process which involved cementing two strips of film together. Despite the apparent difficulties it was used for Cecil B. DeMille's The Ten Commandments

(1922). From 1927, Technicolor superseded the cementing process with dye transfer onto a clear base film. 'Glorious' Technicolor dates from 1932 when a three-strip camera was developed from which three-colour dye transfer prints were made. Becky Sharp (1935) is credited as the first three-strip Technicolor production. Others, including Gone With the Wind, remain impressive, even when printed via other methods, and even when shown on TV.

Needless to say, the Technicolor system was expensive, the camera was bulky, and it was only used for prestigious productions. Colour film as we know it today takes the form of a single film containing three colour-sensitive layers and filters so that the red, green and blue components of the light passing through the lens reach only the appropriate component of the 'tri-pack'. The film is developed directly with no transfer process into a negative image similar to that with which we are familiar from our 35 mm still cameras.

3D

The third dimension has always held an attraction for movie makers, but has proved very difficult to achieve with anything more than novelty value. Truly three-dimensional photography is probably impossible, holograms being the closest alternative which, when done well, do give a convincing illusion of perspective which changes according to the angle of view. So-called 3D movies are probably better described as stereoscopic – two images are shot with the lenses spaced apart so that when they are viewed in the correct way the left eye sees only the left image and the right eye only sees the right image. As in binaural audio, the brain blends the two together to give a moderately convincing illusion. The biggest problem is how to separate the images. In the early days, the two images were projected superimposed on the screen using red and green filters, and were then viewed by the entire audience wearing red and green lensed spectacles. Although this system has been improved, it does not really lend itself to colour. A better way is to project the images through polarizing filters, and this time the audience wears corresponding polarizing glasses. Unfortunately, although quite impressive, the image has more of the appearance of being composed of cardboard cut-outs than being truly solid. Novelty it may be, but it is a fascinating novelty, and worth experiencing at least once in a lifetime. Movies produced in 3D include Hitchcock's Dial M For Murder (1954), It Came From Outer Space (1953) and Jaws 3D (1983).

Why film will prosper

In my view, film will always be superior to electronic images, in terms of absolute picture quality. The reason for this is that the light detectors are so small – on a molecular level – and can capture the fine detail of an image. They fulfil the dual role of storing the image so there is no need to transport a signal from one place to another. I feel it will always be the case with electronic imaging that the detectors and the storage medium will be separated, and therefore the detectors will need to be accessed somehow, and this will inevitably limit how far their size can be reduced. As long as there is a will to do it, whatever improvements that electronic imaging can make will be matched by film. Of course, electronic imaging will take over from film where it is more practical and has more to offer. But for the highest quality, film has it – and will continue to have it for a long time to come.

Figure 13.4 35 mm non-squeezed negative.

Projector aperture
11.3 mm x 21 mm for 1.85:1

Figure 13.5 Release print.

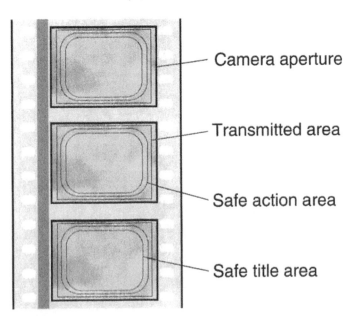

Camera aperture

Transmitted area

Safe action area

Safe title area

Figure 13.6 Television safe areas.

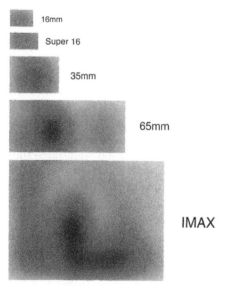

Figure 13.7 Image size comparison.

Original scene

As recorded on film

Unsqueezed
on projection

Figure 13.8 Anamorphosis.

Safe areas

What the camera captures on film and what is later shown on the big screen or on TV may be two very different things. Figure 13.4 shows the SMPTE specification for the image area on film. Notice the gap down the left-hand side which is where the sound track, or rather tracks, are eventually going to end up. The gap is left so that same size copies can be made by simple contact printing. A more recent development, Super 35 mm, spreads the image over the whole available area but then needs to be optically printed through a lens to reduce the image down and make room for the sound track. Figure 13.5 shows the print shown in the cinema which here is masked for a 1.85:1 aspect ratio. The waste of film area that results from trying to achieve a widescreen image with a spherical lens is quite obvious. Figure 13.6 shows a more complex situation which applies when a film is made for television. The camera aperture may be the same as in Figure 13.4, but allowance has to be made for any misalignment between shooting the film and viewing it in the home. Figure 13.6 shows the original camera aperture, the transmitted area, the safe action and safe title areas. These last two are fairly self-explanatory – if you have important action, keep it within the prescribed area. If you want every viewer to see every letter of your title, then keep it within the safe title area. I have only shown a fraction of the number of masking options, and it shouldn't surprise you therefore when you are watching a feature film on television and you see a microphone popping into the top of the picture. The camera operator could see the mic quite clearly in the viewfinder, but it was outside the area intended to the shown in the cinema. Unfortunately the extra height – relatively – of the TV picture has brought a little more of the shot into view. It shouldn't happen, but it is no surprise that it does.

Moving on to Figure 13.7, which film format do you think gives the best picture quality? Outside of the USA it isn't at all uncommon to shoot on bootlace wide Super 16 mm for TV production, whereas 35 mm is very visibly to be preferred. Figure 13.8 gives an example of anamorphosis. The wide proportions of the original are squeezed horizontally by an anamorphic lens to fit it onto the film. The projector would be equipped with a similar anamorphic lens to correct the distortion. You will sometimes notice when a widescreen film is shown on TV that the title sequence is all squashed up. This is obviously so that all the titles are visible. After the titles are over, the anamorphic lens is used to give correct proportions, but the areas at the sides of the image are cropped.

CHAPTER 14

Film stock, film laboratories

I don't think you will be surprised if I say that the film that runs through your 35 mm stills camera is not the same as the film that runs through the motion picture cameras used in top Hollywood productions. The manufacturer's name – Kodak, Agfa or Fuji – may be the same but the stock certainly isn't. The reason we use 35 mm film for stills is simply that it has been the motion picture standard since Edison ordered a stock of film in that gauge from George Eastman for his early experiments. Some years later director Oskar Barnack decided that it would be useful to have a stills camera to test stock before running through his motion picture camera. The first 35 mm stills camera, made by Ernst Leitz, was called the Leica, a name which you may recognize. Since then, however, stills and motion picture films have diverged and, as we shall see, it would not be a good idea to load motion picture film into cassettes and use it for your holiday snaps. It will go through the camera, but will probably totally ruin the processing equipment down at your local photo shop!

Formats

The motion picture film used in the camera is available in three gauges: 16 mm, 35 mm and 65 mm. Sixteen mm film is used for most television drama production in the UK (except where video is used of course) and offers what most viewers seem to accept as a useful image quality, if often slightly grainy and lacking in sharpness. Since the TV safe action area of 16 mm film is a mere 8.4 × 6.3 mm then a certain lack of quality might be expected, no matter how good the film stock. Fortunately, double perforated, or double perf as it is often known, 16 mm film now only accounts for about 5% of the 16 mm market and productions have moved up to single perf or 'Super 16'. Super 16 has the same size perforations but they only run down one side of the film, allowing a larger image area of 12.5 × 7.4 mm. Super 16 was developed in Sweden in the 1970s as a cheaper alternative to 35 mm film for feature film production, but has found its true vocation in television. Double perf 16 mm has a standard aspect ratio of 1.33:1 which is the same as our TV screens, and it has been used for widescreen projection by masking off the top and bottom of the frame. However, since this reduces the image size still further then there is an obvious quality loss. The aspect ratio of Super 16 is around 1.7:1 which comes close to the common 1.85:1 cinema presentation aspect ratio. This makes it practical to shoot in Super 16 for 1.33:1 television, and maintain compatibility with the 16:9 aspect ratio towards which television is hurtling. Obviously when HDTV (High Definition Television) comes of age, the enhanced quality of Super 16 will be essential. In addition to TV presentation, it is also perfectly practical to blow up Super 16 to 35 mm for cinema release, as happened with The Draughtsman's Contract and Truly, Madly, Deeply. It is certainly the TV director's eternal hope that this will happen and his or her work will be shown on the big screen. Although it is practical to work in this way, and audiences will accept it, it is apparent that although 16 mm is adequate for close-ups, long shots do tend to look unsharp compared with 35 mm.

In the USA, TV production budgets are more generous and most are shot on 35 mm. Virtually all feature films are shot on 35 mm too. Even though the image area is about half the size of a 35 mm still camera frame, the fact that the pictures are shown at 24 fps seems to smooth out any grain or possible lack of definition. Since the perceived quality of 35 mm is higher than 16 mm, it is possible to use faster films which do not have such sharpness and low grain and still achieve an excellent result. 35 mm is commonly used for TV commercial production since it provides an excellent original image quality which offers more creative scope to the director than shooting straight onto video.

Going one step up from 35 mm we come to 65 mm, which is the camera origination film used when cost is no object. In the past, 65 mm camera film has been transferred to 70 mm for projection – the extra 5 mm allows space for four magnetic sound tracks – but this type of presentation is rare now. Far and Away (1992) was largely shot on 65 mm, and Little Buddha (1993) had 65 mm segments. If you find yourself deeply trawling your memory to recall exactly when you saw these films in the cinema, it is an indication of how rare 65 mm productions are. Fortunately, it is not a problem to manufacture 65 mm film to order since, in a similar manner to magnetic tape, it is made as a wide roll, slit to width, and then perforated.

Types

As Steven Spielberg's Schindler's List (1993) proved, there is still a market for black and white film in the age of colour. One might have expected that a younger audience brought up on colour would see black and white as archaic, but they have also seen plenty of old black and white news reels and in this case it provided a stunning look of authenticity. 'Look', as you know, is an important word in the film-maker's vocabulary. Black and white film actually makes the Director of Photography's job much harder because he or she only has shades of grey to distinguish the foreground from the background. It is also more difficult to expose since when the eye only has a limited amount of information about the scene, everything that it does receive is of greater importance.

To deal with another minority activity, reversal film was used extensively for television news gathering before the advent of ENG. The advantage of reversal film was that it could be processed into a viewable image straight away with no printing involved. It could also have been given a magnetic stripe for a synchronized sound track. Reversal film is still used, mainly for research applications such as crash tests. Apparently, in this type of work it is important to be able to take measurements directly from the film, which obviously is impossible with video. Additionally, film still has tremendous advantages over video for high speed work. The main drawback of reversal film is its lack of exposure latitude. Since the film that goes through the camera is the end product, then any deficiencies in exposure and colour balance are unlikely to be correctable. The exposure latitude is effectively plus or minus half a stop, which means that you have to get it almost spot on or your film is worthless. Also, since reversal film is in no way optimized for making copies, the film that ran through the camera really is the one and only copy, so distribution on film prints is pretty much out of the question.

Most filming these days is done on colour negative stock. The big advantage of negative over reversal is that there is an exposure latitude of around plus or minus two stops, where each stop is a doubling or halving of light intensity. Although this might seem like an open invitation for the Director of Photography or camera operator to get it wrong, it should be thought of as an opportunity for enhanced creativity. Film can be balanced for the colour temperature of daylight or tungsten illumination. Speed can be measured in EI, or Exposure Index, which relates to the ASA rating used for stills camera films, but is really just a starting point for determining the correct exposure in any particular instance. You might think this is a subtle point, but cinematographers use a very subjective interpretation of exposure, and they will rate the stock they use as they see fit. Some may find that they can give a particular stock a higher rating, given the lighting they are using and the effect they want to achieve. Sometimes a lower rating may be more appropriate. One noted Director of Photography for example places the emphasis more on what people can't see than what they can see, and allows shadows to extend over a large area of the image.

The highest quality film in terms of sharpness and low grain would have an exposure index of 50, which is two stops slower than the 200 ASA film you might use in your stills camera. An exposure index of 50 might be slow in daylight but it is even slower if used with tungsten illumination with the appropriate filter – the exposure index drops to a mere 12! Fine grain is of course vitally important in 16 mm work, but

35 mm users will probably be more likely to take the option of moving to a higher speed. The motion picture camera operator does not have such a wide range of shutter speeds available as a stills photographer, and exposures longer than about a fiftieth of a second are not possible at the normal speed of 24 frames per second. EI 200 stock is very versatile and able to cover a wide range of shooting conditions due to its compromise between sensitivity, grain and sharpness. The fastest general purpose film would have an EI of 500, which is ideal for night scenes or available light without fill. You may think that an exposure index of 500 is still a little on the slow side compared with still films, but as I said, exposure is a subjective thing. In some situations it is possible to underexpose motion picture film considerably, to the point where only the brightest parts of the scene are visible, which in a still photograph may not make any sense at all. But when the brain perceives motion in a very dimly lit scene, it is able to interpret very sparse actual information content into a meaningful image sequence. The same applies to very brightly lit scenes too: what may be totally overexposed and burnt out in a still photograph may be considered artistically acceptable in the movies. It depends on the context. It is, by the way, possible to mix different stocks during the course of shooting a feature film. In fact it is normal, and the different types are designed to intercut well.

For Directors of Photography who want to take the range of brightness a film can accommodate to the limit, modern stock may have a very wide range of latitude. Kodak, for instance, have a stock which they describe as a medium to high speed film with micro-fine grain, very high sharpness and resolving power. The extra latitude is obtained by lowering the contrast, particularly in the lower brightness region, the 'toe of the curve' in technical language. If the film is exposed at the recommended EI 200 then true blacks are reproduced as black, and near blacks are recorded as less dense areas. Exposing at EI 160 results in more blacks with a higher shadow contrast, and EI 250 to 400 gives a less dense black and higher shadow contrast. Contrary to the well established laws of the universe, this film seems to display less grain when it is underexposed. An illusion obviously, but if it looks right then it must be right! Also surprisingly, Kodak rate the film differently depending on whether it is to be used to produce film prints or to go directly into telecine. Due to the shape of the toe of the curve and its interaction with telecine machines, EI 320 to 500 is recommended for this application.

For film shot specifically for telecine transfer, an EI of 640 is feasible. In television shooting, if the film is more sensitive, then lights can be fewer and smaller, set-ups are shorter and location shooting days can be longer with consequent reductions in cost. The spectral sensitivities of the layers are specially attuned to telecine and although prints can be made from this stock, the results are not optimum.

Now for the reason you shouldn't take motion picture film into your local D&P! Manufacturers often supply trial cassettes of film to cinematographers to try out in their stills cameras during the research and development phases of new film types. These cassettes do go out with a stern warning that the film must only be processed by a specialist motion picture laboratory. One reason for this is that the chemical process is different, another is that Kodak motion picture film, for example, has a 'Rem-Jet' backing. This is a black back coating on the base film designed to prevent halation. Halation is where light penetrates the light-sensitive emulsion layers all the way through to the base and reflects back, exposing the emulsions again resulting in a halo effect, hence 'halation'. The black back coating absorbs this unwanted light and additionally protects the film and reduces the effects of static in dry shooting conditions. The Rem-Jet backing is stripped off in the lab.

Intermediate and print film

When a movie can cost up to $200 million or more to make, obviously the actual film that goes through the camera is extraordinarily precious. For this reason it is usually not used to make release prints directly, but goes through an intermediate process. One of the important points about this process is that the contrast of the finished print should be about 1.5, meaning that the contrast is slightly exaggerated from real life. This compares with transparency stock commonly used for stills photography. To ensure that as much information as possible is captured during shooting, and that there is scope for colour grading to optimize the image, the contrast of the origination film will be around 0.6. To achieve a finished contrast of 1.5, it follows that the contrast of the print film should be around 2.5. From the original camera negative, a very small number of interpositives will be made, taking very great care, onto 'intermediate' film. From these interpositives, a larger number of internegatives will be made from which the release prints will be printed. To maintain the correct contrast ratio, the contrast of intermediate film is 1.0. Where the camera film is likely to be on triacetate base film, the intermediates will be on the much tougher Estar base. The print can be on either base, Estar obviously having the advantage of durability, but triacetate having the advantage that if there is a problem during shooting or projection, then the film breaks rather than the equipment!

Laboratories

Just in case you're wondering, motion picture film laboratories are not to be confused with high street processing labs. We all have cameras and take our fair share of holiday shots, and the hundreds of photo counters up and down the land probably get a considerable proportion of their business from us. By and large they do a reasonably good job

– nice bright colours, quick turnaround, moderate prices. Some photo labs even put stickers on our prints telling us exactly what went wrong and that it wasn't their fault! Unless you are a real snapshotter, however, with absolutely no pretensions to photographs of quality, I doubt whether you are entirely satisfied with your D&P. 'How come Auntie Mabel has a greenish tint in her grey hair when she normally has it done mauve?' you ask. 'Why is my white cat sitting on the lawn pink?' 'Why am I so dark in that shot with my back to the sun?' The answers are to be found in the care and attention that is paid to the processing and printing. The chemical part of the processing is largely automated and there is a good deal of assistance available from the film companies to make sure that their products are developed to perfection. The art of converting a negative into a print, however, still calls for human intervention. If left totally to the whim of the machine, the sensors will assess that your cat on a lawn picture is too green overall and therefore the colour balance needs adjusting, so the cat turns pink. Although the negative of you with your back to the sun may have a reasonable amount of detail in your face, even though it is in shadow, the printing machine assesses the exposure over the complete negative and since it is mostly very bright, you turn out very dark. Although there was probably no foreseeable reason why Auntie Mabel's hair might turn out the wrong colour, you simply cannot expect machinery and chemicals to reproduce colours accurately enough to satisfy the human eye all the time. Skilled human intervention is necessary, and more care and attention than our holiday snaps would warrant, or we would be prepared to pay for. Processing of film for cinema and television is in a completely different league to holiday D&P. Far more time is spent making sure that absolutely everything is as perfect as it could be. Let us follow a film through the various stages from shooting to cinema release . . .

Rushes

When you shoot a roll of transparency film in your 35 mm stills camera, it is the film that passes through the camera that will become the actual mounted slide. This is the only copy and if you lose or damage it, that's it. In feature film or TV film production the film that passes through the camera is similarly unique and much more valuable. Although many copies will be made during the production process, the original negative is equivalent to an analogue audio master tape and must be looked after even more carefully since so much expense has gone into its production. After a day's shooting in the studio or on location, the film is taken to a laboratory such as Deluxe Laboratories. Figure 14.1 shows the inside of one of the processors and a few of the many 'soft touch' rollers that the film must pass over while immersed in the chemicals. How did I get the

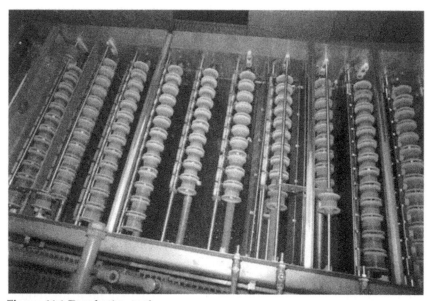

Figure 14.1 Developing tank.

shot without ruining someone's valuable production? Simple – this machine was undergoing cleaning and maintenance with blank film running through. The chemicals are continuously replenished via a tangle of pipe work that would give a studio wireman (or woman) enough nightmares to last a lifetime. As all the film that comes in is processed, it is assembled into laboratory rushes rolls, each roll consisting of a number of scenes adding up to not more than 1000 feet in length – 1000 feet of 35 mm film lasts a little over ten minutes and is a standard unit in the film industry. The original footage will be subjected to 'negative breakdown' which is just the separation of the NG and OK tapes as logged by the camera operator. 'NG' stands for 'No Good'; you know what OK stands for! The Director of Photography by the way, is the person in charge of cinematography and may direct a team of camera operators and assistants. Each roll is numbered in sequence and a laboratory report sheet is made out recording the film title, laboratory roll number and scene numbers. Four copies of the laboratory report are made, one each for the camera operator, the negative breakdown person, the contact person and the editor. Of these four, the laboratory's contact person is one of the unsung heroes of film production. His or her job is to liaise with the camera operator and director throughout the duration of the production from shooting all the way through to show prints. The contact person is the point of contact at the laboratory for the production team, and camera operators will frequently ask for a particular person they know and trust. The contact person will feed back information concerning not only the technical aspects of the process, but also will comment on such matters as whether all is well with the lenses, or whether there is a boom or a bad shadow in shot at any point. Of course, these points will be spotted when the rushes are viewed, but the contact person sees them first and is in a position to give an early warning of any problems.

The rushes – known as 'dailies' in the USA – are working prints of the original negative and will be dispatched to the studio or location for early viewing the next day. For productions that are working to a strict budget the rushes may be printed under 'one light'. This means that the negative is quickly assessed for colour balance and exposure and it is all printed using the same settings. Where there is more money to spend, the rushes will have an initial grading done on a machine like the one in Figure 14.2. The colour grader sees the film on a video monitor, and he has three dials to adjust the red, green and blue components of the image to their correct proportions. Film is very similar to video in the way the full spectrum of light is separated into just the three primary colours to which our eyes are sensitive. The colour grader – the color timer in the States – is very skilled at assessing precise variations of colour. If you have ever tried to adjust the red, green and blue controls on a TV – which are normally hidden under the back cover – you'll know how difficult this is. It's not just difficult to get close to the correct balance, it's difficult to know whether the balance is correct or not. The eyes adapt very quickly to changes in colour balance, and the colour grader has to learn to see colours more

Figure 14.2 Colour grading.

objectively than the rest of us do. The colour grader isn't actually producing the graded print here – he is just doing the assessment and the information he compiles will be passed further down the chain. Modern equipment incorporates computerized assistance, as you might expect. This equipment has a frame store which can hold choice images from several different scenes so that the relative balance can be judged. It would be quite easy to make each scene just a fraction redder than the last, for instance, so that the balance changes throughout the duration of the film. When the grading is finished, the negative is ultrasonically cleaned and the print made using the grader's settings.

Printing

In our amateur efforts at photography we are used to transparencies being transparent, and prints being on paper. In the film industry, a print is a positive copy of the negative. You can see a normal image if you hold the film up to the light, but paper prints are not involved. Figure 14.3 shows a printing machine which is used to copy the negative onto raw

Figure 14.3 Printing.

stock, thus reversing the image. The colour of the light is adjusted at this stage according to the colour grader's recommendations. There are two methods of printing: contact printing and optical printing. Optical printing involves a lens, and an image of the negative is focused onto the unexposed print stock. This gives the option of increasing or reducing the image's size; for example a 16 mm film that was made for TV can be blown up to 35 mm for cinema release. This would of course involve a reduction in quality compared with a 35 mm negative, but the result can still be very acceptable. Also, there are some formats which exploit the available area of the film fully in the camera, and then the image is reduced in the print to allow room for the sound track. Optical printing offers extra versatility but it is quite slow. As in the camera and projector, the film has to be brought to a complete stop on each frame, which of course limits the speed at which it can go through. Contact printing is much quicker since the film can run smoothly and continuously, and a rate of around 1200 feet per minute can be achieved – which is over ten times normal running speed. Of course the printed image must be exactly the same size as the negative. Where ultimate quality is required in the various stages of the production process, optical printing is used. Where a slight reduction can be accepted, the extra speed of contact printing is preferred. The film that is eventually shown in the cinema will almost certainly be contact printed. Figure 14.4 shows how the light from a single bulb is split using dichroic mirrors, and the amount of each colour carefully controlled.

Figure 14.4 Colour mixing.

A further refinement in printing is the use of a wet gate. The gate is the part of the printer where the film is exposed, and in a wet gate printer the negative is immersed in a liquid which will fill in any slight scratches and make them much less noticeable. Of course, this cannot cure deep scratches which affect the coloured layers of the image, causing a coloured scratch, or damage which goes right though to the base of the film. It is better thought of as a technique that gets the best result out of an already good negative. It is very slow though, since the negative has to be allowed time to dry as it emerges from the gate.

Editing and regrading

The film will be edited by a specialist in this fine art working at a location of his or her choice. The work may be done in the traditional film way on a working print – not the original negative! – or it might be done using an Avid or Lightworks nonlinear system. Either way, what is produced during editing is an offline version whose function is to be judged, assessed and modified as necessary, and then put in a can and stored forever, or eventually discarded. The real end product of the editing process is an edit decision list (EDL) which may consist of hand-written frame numbers, or computer-friendly KeyKode markings which link every foot of the edited version back to the original negative. In many ways, the video industry has promoted the idea that film is a bit old-fashioned, but I for one am beginning to realize that this is far from the truth. The OSC/R and Excalibur systems automate KeyKode logging and integrate film feet and frames with video and timecode so a production can pass from film origination through to Betacam SP copies, to nonlinear editing, and back to film again to gain the advantages that each method of working has to offer.

Grading is not a process that happens just once. The film will be regraded and reprinted as many times as necessary, or as many times as the budget can stand, until a 'cutting copy' can be made which is edited and graded to the highest standard achievable at this point. Final grading is done by eye rather than on the machine described earlier. The director, contact person and colour grader will now get together and discuss every aspect of colour control and density, considering not just the individual scenes, but the changes from one scene to the next. Regrading and reprinting continue until the work is perfect and an 'answer print' is produced which the customer – the production company – can approve. Apparently everyone wants to take part in the colour-grading process and it can become a matter of grading by committee, which may or may not always produce the best results. Perception of colour is a very subjective matter, and even high status actors and actresses may want their say (and some even know what

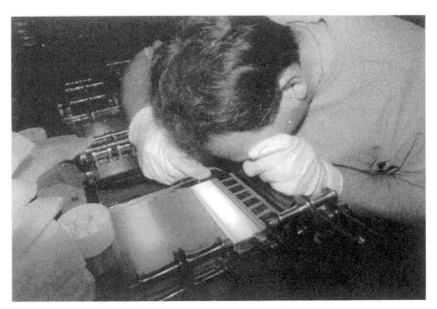

Figure 14.5 Negative cutting.

they are talking about!). Figure 14.5 shows the negative cutter at work. This is not a creative process, but it does require great care and attention – and nerves of steel probably, considering the value of the material being cut.

Release print

When the answer print is finally and irrevocably approved, the laboratory starts on the manufacture of release prints. If the production is for television, only two or three prints may be required. For worldwide release of a major feature, however, the quantities may rise to 5000 or more. The original negative is irreplaceable and the insurers would object strongly to running it through even the best printer thousands of times. Also, the colour grading information may be so complex that it would be impractical to make prints at high speed. Only a few prints will therefore be made from the original negative and you will probably have to go to Leicester Square to see the very highest quality that film is capable of. The vast majority of prints will be made using an 'intermediate' copy: a print is made from the original negative on negative film stock which produces a positive image, but with flattened contrast which retains more of the information in the negative than an ordinary print would. From this

Figure 14.6 Checking the release print.

interpositive, an internegative is made which becomes the master printing facility. Should this become damaged or worn, sections can be replaced by reprinting from the interpositive while the original negative stays safe in its secure storage. The machines on which release prints are made need to be efficient, so the film runs forwards then backwards without rewinding. There is also a 'longer length' printer which can handle a whole feature film on a reel up to 18 000 feet long and about 6 feet in diameter. Figure 14.6 shows the final checking of the release prints where the operators get to see the latest films long before we normal people get the chance. I don't suppose they go to the cinema very much!

Once distributed, the release prints will last for a considerable period of time when handled by experienced projectionists, although you will notice that the beginning and end of each reel is inclined to gather dirt and scratches during use. Inexperienced projectionists can apparently destroy a print on its first showing. They may seriously damage the projector, too, since most prints are made on polyester material these days which just does not break. The older acetate prints would tear and protect the projector, which can be seen as an advantage, but apparently consideration must be paid these days to disposing of the print when it has reached the end of its useful life because no-one wants to see it any more, and polyester is capable of being recycled. An ignominious end for the stars!

Special processing

Film manufacturers issue precise recommendations on processing to achieve the best results from their products, but once in a while it is necessary to bend the rules a little. One fairly common technique in photography, both still and motion picture, is forced development or 'pushing'. This is where the camera operator deliberately underexposes the film and asks the laboratory to leave it in the developer for longer than normal to compensate. This might be necessary if the light was insufficient for normal exposure and high speed stock was not at hand, or someone might have made a mistake! At up to one stop of underexposure – a halving of the light reaching the film – forced developed scenes can be intercut with normally processed material with few noticeable side-effects. At two or more stops of pushing, the grain of the film becomes emphasized and the fog level rises. Beyond this is uncharted territory where the dreaded 'crossed curves' phenomenon may occur. In normal processing, each of the three colour-sensitive layers responses very evenly to low, medium and high levels of light, and all points in between. With forced development, however, the red layer may become more sensitive than the blue at low light levels, and less sensitive than the blue at high light levels. This will produce a negative that is impossible to grade correctly. Of course, this is sometimes what the director and camera operator want for 'artistic' effect.

Another special processing technique, although not common, is 'silver enhancement' as used a long time ago in Moby Dick (1956). The light-sensitive component of unexposed film is a compound of silver, which is used to form dye images and is then bleached out. In the silver enhancement process, the silver is allowed to remain in the film and has the effect of making the colours less saturated, becoming muted and less 'picture-postcard-like'. Interrupting the normal flow of work in the laboratory pushes up the cost of course, which is why this effect is not seen more often, but it is still there for when an enterprising director chooses to use it.

A third technique, which is not really processing but is worth a brief mention, is the production of separation positive prints. This is where the three primary colours are separated out onto three individual films. These are most commonly used for the production of special effects, but one interesting point is that the dyes used in colour films will inevitably fade in time. Separation positives are made onto black and white film stock where the image is made from metallic silver which cannot fade. The moral is that if you consider your film to be a work of art and want it to be viewable in the next century and beyond then you had better have separation positives made, otherwise it may not stand the test of time in more ways than one.

Video tape

When we look at modern video equipment, it is easy to recognize the amount of research, development and sheer engineering brilliance that has gone into its design and manufacture. We can readily appreciate the improvements in picture and sound quality made available by the latest digital video recorders, and we can marvel at the comparison with earlier models developed over the last thirty years or so. However, it isn't quite as easy to appreciate the similar developments that have occurred in video tape, and audio tape as well for that matter. We are all very familiar with audio tape and we expect the latest development from one of the major manufacturers to offer a small and only marginally audible improvement over the previous generation. But when all the developments over a period of years are aggregated, the work of the chemists, physicists and production engineers is quite obvious. Older audio tapes had low output and a rough, grainy noise characteristic. Modern tapes have a much higher output, and what noise remains is much smoother and less obtrusive. Video tape has developed over the years too, but as well as providing increments in picture and sound quality, new developments in tape have resulted in new video formats. As tape science has improved, video recording has moved from fuzzy pictures on two-inch reel-to-reel tape back in 1956 to crystal clear images on half inch tape today, neatly packaged in a handy cassette with four channels of twenty bit digital audio thrown in for good measure.

History of tape

The earliest recordings, audio recordings obviously, were made on wire, similar to piano wire, or on hard-rolled steel tape. In the 1930s metal alloys were tried, the most successful being Vicalloy which combined iron, cobalt and vanadium. In Germany, the IG Farben company had developed a fine powder of carbonyl iron for which they were looking for an application. They chanced upon the idea of coating it onto a paper ribbon to make something we would now recognize as tape. Unfortunately, of all the possible magnetic materials they could have chosen, carbonyl iron had particularly poor magnetic properties, but the tape was cheap and easy to splice – a procedure that understandably was not too successful with steel tapes and wires. The natural course of progress soon led to tapes using magnetite and gamma ferric oxide, but the science of using these materials was still at an early stage and the tapes had low coercivity (more on this later) and high print-through.

The 3M company was responsible for what we would recognize as the first modern recording tape. 3M (which stands for Minnesota Mining and Manufacturing) was then, and still is, a major manufacturing company and their principal concern was in coating various materials onto a substrate of some kind. They were able to coat acicular (needle-shaped) particles of gamma ferric oxide onto a cellulose acetate base. The particles were aligned along the length of the tape and the result was what was considered at the time to be an excellent recorded sound quality.

Cellulose acetate was adequate as a base material for audio tapes, and provided a smooth surface, but unfortunately its strength was not great and any edge damage to the tape would soon result in a break. In Europe, polyvinyl chloride (PVC) was preferred, but this had the disadvantage that rather than breaking cleanly it would stretch. Obviously a break can be joined back together but stretched tape is ruined forever. Neither of these materials was suitable for the rigours of video tape recording, but fortunately DuPont had recently developed the polyester material Mylar which had the right degree of strength and elasticity, although there were initial problems with surface smoothness and coating adhesion.

Further developments in tape technology include chromium dioxide coatings, cobalt modified oxides, metal particles and evaporated metal film.

Magnetic properties

The two key properties of magnetic materials for our purposes are coercivity and retentivity. Coercivity is a measure of the amount of magnetic force necessary to magnetize a material, and is also therefore a measure of how difficult the material is to demagnetize. Magnetic materials can be classified according to their hardness or softness. A hard magnetic material is difficult to magnetize but retains its magnetism and is a permanent magnet. A soft magnetic material is easy to magnetize, but loses its magnetism readily – tape heads are made from soft magnetic material. Retentivity is a measure of how much magnetism a material retains as a result of exposure to a magnetizing force.

The net result of all the magnetic theory this involves is that good retentivity equals good low frequency performance, good coercivity equals good high frequency performance. Retentivities and coercivities have been steadily rising over the years, but the main thrust of development has been towards higher coercivity since this leads to the possibility of shorter recorded wavelengths and therefore better packing

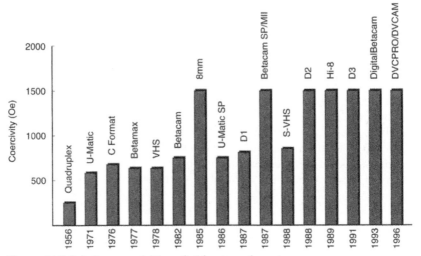

Figure 14.7 Relative coercivities of video tape formats.

density. Coercivity is measured in oersteds and has risen from 10 Oe for steel wire or tape at the beginning of the century through 250 Oe for early post-war tapes, to 1500 Oe or more for modern metal particle tape. Figure 14.7 shows the progression for the various video tape formats.

One technique used to increase the coercivity of the magnetic coating is to increase the magnetic anisotropy of the particles. Small magnetic particles will have a preferred direction of magnetization, the direction in which it is easy to magnetize the particle. This is called the easy axis, and conversely there is a hard axis where it is difficult to magnetize the particle. Where a particle is forced to reverse the direction of its magnetization, the direction of magnetization will have to pass through the hard axis and the greater the anisotropy – the difference between easy and hard – the higher the coercive force that will be required. Anisotropy can be increased by using needle-shaped particles or by adding cobalt to the lattice of gamma ferric oxide particles. TDK's Super Avilyn technique uses gamma ferric oxide particles which are coated with a thin layer of cobaltous oxide. The latest generations of tape use pure metal particles (rather than oxides) and, to a lesser extent, evaporated metal film. The use of metal particles has been explored since the 1930s, but their inherent instability was not overcome until the late 1970s.

Physical properties

Magnetic tape consists of three layers: the base material (thickness between 10 µm and 20 µm) which is magnetically and electrically inert, the magnetic coating (3 µm to 6 µm) and back coating (0.5 µm to 1 µm). Experiments where the base material was itself made magnetic proved fruitless since the more magnetic material that was incorporated into the base, the weaker it became. As important as the magnetic particles themselves is the component of the tape known as the binder. The binder sticks the particles together and holds them firmly onto the base material. The binder has a number of constituents, each with a particular function:

- Materials which are suitable binders may not in themselves be sufficiently flexible so a plasticizer will be added to correct this.
- Obviously the magnetic particles in the coating should be evenly spread over the surface, so wetting agents are added to the binder to facilitate this.
- When the magnetic coating is applied to the base it is important that it should flow correctly so special flow agents are added to the binder mix.
- Lubricants are added to reduce friction against the tape heads and there will also be head cleaning agents.

Since the tape is going to spend a large part of its life being wound at high speed, and since the base material is a very good electrical insulator, there is a high likelihood of static charges developing, so the binder will have an antistatic agent to combat this. The back coating is also conductive for the same reason.

Manufacturing

Manufacturing starts with large rolls of base material to which the magnetic coating is applied. Figure 14.8 shows the manufacturing procedure in basic detail. The first stage is to prepare the magnetic particles in a sandmill, and then to combine these with the ingredients of the binder to form the dispersion. This is coated onto the base film, and before the dispersion is dried, the particles are magnetically aligned along the length of the tape. Drying is done using air flotation to maintain the initial surface quality, which is further improved by a process known as calendering. The jumbo is rolled up and allowed to cure before going on to the next stage. While the jumbo is being processed, sophisticated equipment monitors the coatings to ensure

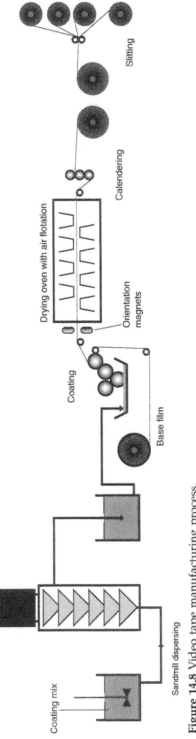

Figure 14.8 Video tape manufacturing process.

their accuracy. At each stage the operators perform quality checks before the jumbo moves on. After they have dried and cured, the jumbos are slit into half-inch widths with rotary blades and wound onto large reels.

Automated machinery takes the large reels of tape onto the next stage, which is loading into cassettes. The cassette loaders are fiendishly clever pieces of apparatus which take an empty cassette, cut the leader tape which is already inside, splice on the video tape and wind it into the cassette, and then splice the other end onto the other half of the leader. When one large reel of tape is finished, the machine automatically changes over to another and flashes a light to alert the operator. Finished cassettes go on to another room with automated packing and wrapping machines.

CHAPTER 15

Cinema technology

Projection technology is probably the longest established of any of the fields of film and video. A 1930s projectionist time-warped into a present-day theatre would recognize the equipment and would be able to operate it almost straight away with very little assistance. Indeed, they might find absolutely no difficulty at all because the projector actually dates from the 1930s, although it will almost certainly have been modified to take account of the occasional development in projection technology since then.

As you know, motion picture film consists of a sequence of still images which are projected at a rate of 24 frames per second. The film runs at this speed, not because of any visual requirements but because the early sound films demanded the equivalent of 18 inches per second to achieve a reasonable degree of intelligibility. Previously, the standard silent movie speed had been 16 fps, and audiences of the time seemed satisfied with that. These days, 24 fps is seen as just fast enough to achieve smooth motion, although camera pans will give the game away and a higher rate certainly could be wished for. It will probably come in time. Also as you know, 24 fps is not a fast enough rate to eliminate flicker. The eye requires a flicker rate of at least 48 Hz so that the effect of persistence of vision joins the images together into what appears to be a steady level of brightness. The function of a projector therefore is to pull down a frame of film, stop it dead in its tracks and project it twice on the screen with a brief dark interval in between, then pull down the next frame and do the same. Repeat this prescription twenty-four times a second and you have motion pictures!

The basic mechanism of the projector is known as the Geneva movement, which was invented in the 1890s and is still in use today. There are other methods: simpler mechanisms are used in equipment used for viewing dailies, and of course the IMAX projector requires an altogether gentler film handling system due to the sheer physical size of the frame. The Geneva movement is brilliant in its simplicity and achieves a high degree of reliability. It incorporates a cam, a pin and a Maltese cross with four slots, as shown in Figure 15.1. The circular cam rotates in an eccentric fashion once per frame. For three-quarters of a turn,

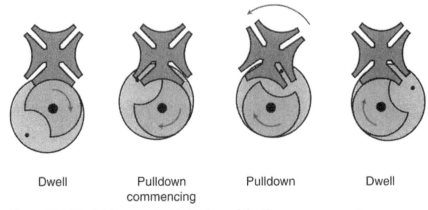

Dwell   Pulldown   Pulldown   Dwell
               commencing

Figure 15.1 The Maltese cross mechanism of the Geneva movement.

Figure 15.2 Simplex PR1060 projector with the 5-Star Soundhead.

the Maltese cross simply stays where it is and the surfaces just rub together. During the fourth quarter turn the curve of the cam gives way to the pin which is positioned to engage into one of the slots of the Maltese cross. This causes a quarter-turn rotation of the Maltese cross, which in turn drives a shaft connected to a sprocket wheel. The wheel will pull down precisely four sprocket holes of the film, therefore advancing it by one frame. As the motion repeats, the film will stay steady for the 'dwell' phases of the Geneva movement, and advance again during the 'pulldown' phase. Thus, the film moves in an intermittent motion, staying rock steady for three-quarters of each turn of the cam, and being pulled down for the fourth quarter turn. Interestingly, it is possible to fit a dual sprocket with sixteen teeth for 35 mm projection and, spaced further apart, twenty teeth for 70 mm projection. The Geneva movement itself requires no modification.

To provide the necessary 48 Hz flicker, a shutter is directly linked to the Geneva movement. The shutter has in the past been made with two blades that cut out the light once as the film is pulled down (or the audience would see the blurring of the image), and again during the dwell phase of the movement. The problem with a blade shutter is that to

Figure 15.3 Century MSC/SA-TU projector with the W/R3 Soundhead.

obscure the image completely, it has to traverse the entire height of the frame. This takes time and valuable light energy is lost. Better than the blade shutter is the drum or cylindrical shutter which cuts the light starting from the top and bottom of the frame simultaneously, quickly meeting in the middle. The timing of the shutter has to be adjusted accurately so that it coincides exactly with the movement of the Maltese cross mechanism (in modern projectors, this can be done while the film is running).

Lamphouse

Projectors have traditionally been of modular design. This probably stems from the efforts of theatre owners and projectionists of an earlier era who were keen to take advantage of any improvement in any part of the exhibition process (including – rumour has it – turning up the theatre's heating to enhance the sales of refreshments). The projector can be broken down into five main sections: lamphouse, the projector itself, reels or platters, lens and sound head. The function of the lamphouse is obvious, but improvements in technology have led to very significant changes in theatre design, eventually to the multiplex theatre as we know it today. Early projectors used a carbon arc light source since this was the most brilliant form of illumination available. A carbon arc consists simply of two carbon electrodes, incorporating metallic compounds and coatings, the tips of which are brought close enough together that when an electric current is applied, an arc is formed similar to that formed by electric welding equipment. Note that carbon arc light is not to be confused with limelight where a block of calcium oxide (lime) is heated with an oxyhydrogen flame – quite a different thing. The electrodes in a carbon arc lamp could be up to 16 mm in diameter taking a current of 250 amperes at 120 volts d.c. (yes, two hundred and fifty amperes) giving a power of 30 kW. This of course would be for quite a large theatre, and would give a light of good quality at a colour temperature of 5500 K. The crater in which the arc is formed would be reflected by a curved mirror to focus the light into condenser lenses which further focus it onto the film in the gate. The problem with carbon arc lighting is that the electrodes burn away. Technology of course was able to provide motorized electrode transport so that the operator did not have to make continuous manual adjustments, but electrode replacement always was a task that had to be done by hand. Since even the longest electrodes lasted only about an hour, it was impossible to show a full length feature on one projector, hence two had to be used with a changeover between reels.

During the 1950s, however, xenon arc bulbs became available which would give an operating life of 1000 to 1500 hours. In the early days, xenon bulbs were only available up to 2 kW, which is only enough for a small theatre, but now bulbs of up to 7 kW and more are available which have the efficiency to light the largest screens in use. With a xenon-equipped lamphouse, there is no reason why a single projector cannot be used for an entire feature. The operator will monitor the hours a bulb has been in use for and change it ahead of failure when it reaches its rated lifespan, or perhaps 125% of its rating since this is likely to be a conservative estimate. A modern 1 kW bulb would be guaranteed to last 3000 hours, a 7 kW bulb for a mere 500 hours. It is now perhaps only of historical relevance but there is a certain risk that when a xenon lamp fails, it will fail catastrophically and possibly damage the lamphouse, necessitating resilvering of the mirror. For this reason, projectors are not equipped with automatic changeover, which might otherwise be expected.

The efficiency of the lighting depends on a number of factors. The reflector and condenser lenses have to be adjusted to get the maximum illumination into the gate. The size of the screen and whether it is curved or not are significant factors, as is the degree to which it is perforated to allow sound to pass through. Ideally the screen brightness should be between 12 to 16 foot lamberts at the centre of the screen, hopefully not fading too noticeably towards the edges.

Reels and platters

Although the notion of a 'reel' of film being about 1000 feet and lasting around ten minutes is prevalent in film production, for projection a reel has traditionally been twice this length, 2000 feet. Since this only lasts around twenty minutes, obviously several reels are necessary to make up a complete feature film. As explained earlier, the original carbon arc lamps could not last the full duration of a feature film so multiple projectors were necessary for that reason also. The reason why a reel was limited to 2000 feet until the 1950s was that prints were made on cellulose nitrate base material. So-called nitrate stock is highly flammable and the combination of the intense heat in the gate and an flammable film was a significant risk (in 1928 the Gaiety Theatre in Courtenay, BC, was destroyed by fire when a nitrate-based film called 'Safety First' burst into flames). To reduce the risk of a fire in the gate spreading to the reels, the reels were limited to 2000 feet and surrounded by fireproof enclosures, the entrances to which were sealed by tight-fitting fire-trap rollers. When nitrate film gave way to non-flammable acetate and polyester, the spool boxes were retained as protection against dust and dirt, but the spool size was increased so that the projectionist only had to perform perhaps one changeover during the feature.

The ultimate conclusion of the increase in reel duration is to have the entire feature on one reel which might be up to 54 inches in diameter, holding up to four and a half hours of film. A reel of this size is much too big to be mounted vertically so a system of horizontal platters has replaced the traditional vertical reel almost universally. Films are normally supplied to theatres on 2000 foot reels, and they will be 'made up' on the platter, in other words spliced together so that they can be shown as one continuous reel. There are always at least two platters. One is the supply platter where the film is released from the centre. On the other platter, the film is wound up again starting – of course – from the centre. The result is that when the entire film has been shown, the reel is ready to go again straight away without rewinding. Why don't audio and video recorders work like that? Usually there will be a third platter which is used as a make-up table so that while one film is showing, the projectionist can be making up the next feature scheduled to show in the theatre. It is likely that 2000 foot reels will eventually be replaced by Extended Length Reels (ELR) capable of holding 6800 feet of acetate or 8000 feet of polyester film. This will reduce the time taken to make up a film on the platter and should also reduce the possibility of damage during these operations. (The film has to be 'unmade' to return it to the distributor. Splices between reels are made with yellow striped 'zebra' tape to allow fast location.)

Once the platter system was in place it was only a small step to employ one set of platters with two or more projectors, thus serving two or more

Figure 15.4 Platter system as commonly employed in multiplex theatres.

theatres of a multiplex with the same feature. Once upon a time a projectionist was kept fully occupied running one theatre (apparently you can always recognize a projectionist by their ability to finish a hamburger in three bites). Now that same projectionist will still be busy but serving several audiences with different films. There may still be the occasional problem, but personally I can only remember one occasion when the film ground to a halt (it was during The Exorcist!). 2000 foot reels and dual projectors are still in use, but only for presentations where the film is going to be shown once, or just a few times, otherwise the platter system wins all ways in terms of efficiency.

Lenses

The requirements of a projection lens differ from camera lenses in a number of respects. Firstly, there is no need for a variable iris as the lens will always work at full aperture. Associated with this, the lens should display its best performance at full aperture, typically f1.8 or better, whereas a camera lens will be designed for best performance around the centre of its aperture range. The lens will be designed to compensate if necessary for the curvature of the film in the gate. Some projectors hold the film flat in the gate, some curve the film to hold it more rigidly and if this curve were not compensated for, parts of the image would be out of focus. Also, the lens must display consistent focus on the screen, whether the screen is flat or curved. All elements of the system must be in alignment for optimum performance. The focal length of the lens may be from 75 mm in a small theatre up to around 125 mm in a larger theatre.

Many films are shot in an anamorphic format where the horizontal dimension of the image is squeezed by the camera lens in anticipation of being stretched out again on projection. This requires either an anamorphic lens, or an anamorphic attachment. Designing an anamorphic lens for projection is an easier matter than for a camera since the object-to-image distance is more consistent. Shooting anamorphic close-ups has in the past caused problems such as 'anamorphic mumps' where actors' faces seem strangely distended horizontally.

Since films are shot in a variety of formats, including anamorphic and non-anamorphic, it is impossible to equip a projector with a single lens that will suit all. Hence several lenses are mounted on a 'turret' and the appropriate one rotated into position for the presentation.

Sound head

Perhaps the greatest recent technological developments in projection have been in sound. In the early days, silent projectors were upgraded by fitting a sound head and associated components. Modern projectors are fitted with optical sound readers as standard, but the development of digital sound systems for the cinema has meant that further adaptation has taken place and projectors may sport a combination of Dolby, DTS and SDDS sound readers as the theatre owners find necessary. If you look at a projector you may see vacant holes where the Kodak CDS (Cinema Digital Sound) system has been removed and placed in storage in case there is ever a need to revive Warren Beatty's Dick Tracy (1990)! The requirements of sound, sticking to optical analogue sound for the moment, are vastly different to the requirements of image projection. As explained earlier, projection requires a stop-start intermittent motion to suit the sampled nature of the moving image. Sound reproduction, however, requires a smooth continuous motion or massive wow, flutter and modulation noise will result. Smoothing rollers damped by heavy flywheels are therefore fitted between the gate and the sound head to even out the motion. Since the image and sound are picked up at different places along the length of the film, there has to be a standardized 'sound advance' so that the two will be in sync. Twenty-one frames are allowed between sound and picture to provide the necessary advance.

In the traditional optical sound system, which every sound projector still possesses, a small exciter lamp is focused on the sound track area, the varying width of which attenuates the light in response to changes in the sound waveform. This is picked up by a photoelectric cell and amplified in the normal way.

Cinema systems

These days, the projection booth of a cinema or multiplex is an entire system rather than a collection of components. The projector will be mounted into a console onto which the lamphouse is bolted and into which the sound system and other electronics can be installed. Much of the presentation process can now be automated. A computer-controlled automation system will perform routine tasks previously done by the projectionist, allowing fewer projectionists to operate more screens. The automation system will be capable of controlling the projector, lamphouse, picture and sound changeover between projectors if necessary, lens turret and aperture masking to accommodate different formats, and also the curtains and dimming of the auditorium lighting. The system will stop automatically and an alarm will sound if there is a break in the film so that the projectionist can intervene and fix the problem. If the worst comes to the worst, then manual operation is still a possibility.

Although projection technology goes back a long way into the past, continual incremental improvements in technology have allowed more efficient operation, and hopefully a more efficient means of entertainment, keeping more theatres open during the lean years and keeping ticket prices down. Soon film projection will meet challenges from digital technology, but the movie theatre experience as we know it still has plenty of time left to run and run.

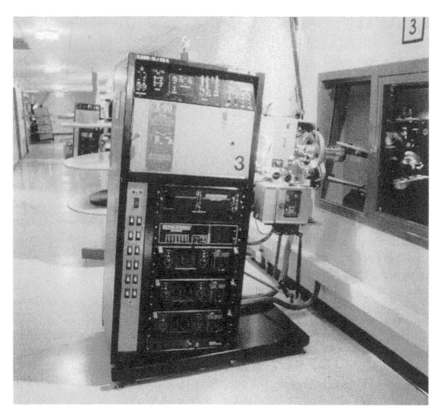

Figure 15.5 Projection console including lamphouse, projector and ancillary equipment.

Changeover

Where films are presented using dual projectors, then the projectionist has to know when to change over from one projector to the other. The two projectors are loaded before the presentation starts with the first and second reels of the movie. Just before the first reel comes to an end, the projectionist watches the screen for the first cue mark – a dot in the top right corner of the image – upon which he will start the second projector via the changeover controller, also switching on the lamp at this time. Both lamps will then remain on for the whole show to avoid thermal stress. The second cue mark indicates that the reel will end in 20 frames and the projectionist must press another button which will switch the soundtrack and also open a changeover shutter on the second projector while closing that on the first. This process will be repeated for every change of reel (and let's hope they are in the right order!).

CHAPTER 16

IMAX

If it's a stunning visual experience you're after, and you don't actually want to go there yourself, then you can't beat IMAX – the largest motion picture film experience ever developed. In IMAX cinemas, the audience may be warned before the presentation that if they should feel the onset of vertigo, they should simply close their eyes until it subsides. This is no idle warning. The IMAX experience can be a spectacle on the scale of the biggest you will ever see, and as thrilling as a theme park ride without ever leaving the comfort of your cinema seat.

To trace the development of IMAX technology we have to look back over thirty years to Expo '67 in Montreal, Canada. Three people who would later become the founders of Imax Corporation were present at a show in which multi-image projection was almost a key theme. But it was done in a clumsy way with multiple projectors and, as had previously been found with the original three-projector Cinerama format, it is virtually impossible to screen multiple images without the seams showing. Graeme Ferguson and Robert Kerr were showing a film they had produced, Polar Life, at Expo '67. Roman Kroitor was there too with the multiple image experimental film Labyrinth. The three had in common a desire to make movies in a bigger and altogether better way and founded the company that became Imax. Imax was subsequently approached by a delegation from Japan making plans for Expo '70 in Osaka. They promised to deliver to Expo '70 not only a film, but a whole new system of shooting, production and projection.

Large-format production requires large-format film. Multiple images had already been ruled out and it was considered impractical to attempt to develop a new format absolutely from scratch. However, there had already been a good deal of production done in the 65 mm / 70 mm format – 65 mm for shooting, 70 mm for projection – and stock was available right out of the Kodak catalogue. 65 mm film is employed in various conventional formats including Todd-AO, Super and Ultra Panavision, but the image size on the film is restricted by the available width between the sprocket holes to 48.5 mm. Imax realized that just as the VistaVision format turned 35 mm film sideways to allow a wider image, 65 mm could be turned sideways too, so that the width of the image could be allowed to extend over 15 sprocket holes to 69.6 mm. The area of a large 15-perf frame is therefore some 3376 mm^2, which is more than three times the area of a conventional 5-perf 65 mm frame, and more than ten times the area of 35 mm. Comparisons with 16 mm are, I feel, rather superfluous and video – even digital video – doesn't even enter the arena. One would be entitled to ask why the 15-perf width was chosen rather than a greater or lesser figure. Apparently it had to do with the manufacturing process in which the perforations of the film display cyclical slight irregularities. 15-perf works, but increasing the width further apparently did not work so well. Also, the 15-perf width of 69.6 mm gives an aspect ratio a little bit wider than 4:3 which corresponds well with the human field of vision in the vertical and horizontal directions. If earlier large image films had used widescreen to trick the eye into thinking it was seeing something spectacular, IMAX presentations didn't need it – it was just huge! It is probably fair to say that IMAX production accounts for the bulk of 65 mm and 70 mm stock sold right now, conventional 65 mm production currently being something of a rarity.

The camera

Designing a camera to handle the large format turned out to be not so much of a problem for Jan Jacobson, a Norwegian working in Denmark, who simply turned the film on its side and used a more-or-less conventional mechanism scaled up in proportion. This part of the project took a mere two months. Even though the format is much larger and the rate of travel of the film is faster, at least the negative only has to go through the camera once, meaning that as long as the film isn't actually damaged, a little wear and tear does not make much significant difference (it does in the projector where the film must be shown many times). But there were certain problems: the larger film format took longer to shift from one frame to the next, therefore the shutter angle (analogous to shutter speed in a stills camera) was smaller than on a conventional movie camera. This meant that more light was necessary to expose the film adequately, or a more sensitive film had to be used. But in the process of the format's maturation, this has been solved by careful choice of materials and refinements in the design. For instance, the claw arm is now made of beryllium which is a very stiff material and very suited to the task. These days it is even possible to build high speed IMAX cameras for slow motion cinematography with frame rates up to an amazing 96 frames per second! Since large format film running at 24 fps is consumed at a rate of 168 centimetres per second (66 ips), 96 fps operation achieves a speed of 672 cm/s or 264 ips. The latest IMAX cameras running at normal speed can achieve a shutter angle of 180 degrees, which means that the shutter is open for half of the duration of a single frame. This figure is comparable with conventional 35 mm and 65 mm cameras.

Figure 16.1 IMAX camera. (©Imax Ltd 1999.)

A standard lens for IMAX production is of 40–50 mm focal length which, considering the size of the format, makes it a wide angle lens. IMAX is of course a 'vista' medium rather than a close focus medium – that is its strength – so a lens that can capture the world in its full width and height is appropriate. Where necessary of course other lenses can be employed, down to a 30 mm 'fisheye' which gives an extraordinarily wide angle of view. (The crew have to be careful where they stand!) The lenses are not particularly unusual in any way and, apart from the specially designed housing, you could easily come across them on a wedding photographer's Hasselblad. The laws of the natural universe unfortunately apply and the larger the image size, the less the depth of field. Focusing therefore has to be absolutely spot on with very little margin for error. In the early days of Imax when the shutter angle was not as wide as it is now, and the film received a shorter exposure, this was something of problem as opening up the aperture to compensate reduces the depth of field further. Fortunately, the wide angle lenses commonly used in IMAX production intrinsically have a better depth of field than the longer focal length lenses used in conventional film production, which is a fortuitous advantage that still applies.

Since the 15/70 format is so large, one might be tempted to imagine that the camera is correspondingly large and heavy. In fact, IMAX

Figure 16.2 IMAX on Everest. (©Imax Ltd 1999.)

cameras are surprisingly compact, typically around 30 kg (70 lb) in weight. This is not exactly lightweight, but it is certainly compact enough for the helicopter shots that seem to be one of the mainstays of large format movies. IMAX cameras have been taken on board the Space Shuttle and the results are truly spectacular (probably guaranteeing NASA's budget for the foreseeable future). The lightest IMAX camera at just under 17 kg (37 lb) was the one that was taken to Mt Everest. Lightweight though it might have been in comparison with the others in Imax Corporation's rental stock, I'm not sure that I would have cared to have it in my rucksack! Around this weight, IMAX cameras are also usable with Steadicam, so the camera operator can keep up with the action, although the Steadicam system itself adds considerable extra weight to the total that the poor operator has to carry. Surprisingly, IMAX cameras are seldom blimped so, considering the noise made by the rapid film transport, sync sound shooting is not often undertaken. In fact the IMAX camera is more likely to be put in an underwater housing. Perhaps the underwater housing might itself make a suitable blimp providing at least a few decibels of camera noise reduction.

Projection

Although the IMAX camera was developed in a short period, the projector proved to be a much bigger problem. A conventional projector with an intermittent stop-start mechanism was simply not able to shift the film fast enough. The acceleration and deceleration involved in projecting conventional 70 mm film is close to the limit, but to do that over a 15-perf frame is impossible without damaging the film. Fortunately, around the time of inception of the IMAX project an Australian engineer named Ron Jones was working, through his fascination with motion picture equipment, on an alternative method of projection for smaller formats. Rather than using the brute force traditional technique he devised a method where the film would wind onto a drum where a little bit of slack would be created which would travel round the drum, eventually landing on fixed registration pins for the projection of one frame. Jets of compressed air guided and cushioned the movement of the film so that it was much gentler than conventional projectors in that the perforations are not used to pull down the film (or pull across, as it would be in the IMAX projector). They are only used to register the film on the pins. The wavelike motion ensures that each frame is lifted gently from the pins and the next is laid down just as gently. At least this is the idea, since it didn't work satisfactorily straight away, as might be guessed. This system was called the 'Rolling Loop' and patented by Jones.

Jones' original 35 mm rolling loop projector worked well enough to prove the concept and Imax Corporation bought the patent.

Figure 16.3 IMAX projector. (©Imax Ltd 1999.)

Unfortunately, when the system was scaled up to 15/70 dimensions, although the film could be projected slowly, when it was attempted to crank it up to 24 frames per second the system displayed what was known as the 'autumn leaf effect' as the film was shredded. A partial solution to this was to employ extra moving registration pins which were cam driven to slow the film down gently as it landed on the fixed pins. This allowed projection speeds up to 18 frames per second, still not fast enough. The complete solution lay in a larger drum which ultimately allowed the required 24 fps rate, and is now capable of very much faster speeds, as we shall see.

To ensure complete image stabilization the frame being projected is held against what is effectively the rear element of the lens by a vacuum. This element is curved to correspond to the curvature of field of the lens so that the projected image is in precise focus from corner to corner. To provide a bright screen image a large and powerful lamp is used. How powerful it is would stagger the imagination of a mere sound person who would think that a Sony 3348 drew a lot of current. The lamp in an IMAX projector consumes up to 15 kW of power (five times the largest allowable domestic electric heater). Of this 13 kW is wasted as heat, a further 1 kW manifests itself as non-visible light and 1 kW finally makes its way to the 69.6×48.5 mm frame. The unwanted portion of the lamp's output is cooled by water, and I imagine an IMAX theatre could lay on hot showers for the staff without any additional electricity bill. As an interesting comparison, the lamp for a lighthouse would be of the order of 1 kW to warn off shipping miles away at sea. Obviously, light of this intensity takes its toll on the print, and the image will be subject to fading eventually. However, dust and scratches will render the print unusable after around 1500–2000 passes, although the perforations will last for over 5000 passes without undue wear. Since the whole point of the IMAX experience is the ultimate in image quality, it does make sense to replace a print sooner rather than later.

The IMAX cinema

The IMAX theatre has to be purpose-built for the format. A conventional cinema will not suffice. The IMAX screen is enormous, often stated as 'more than five storeys high'. This translates as up to 27 m high by 36 m wide. This is large enough to show certain species of whale life-size. The idea behind the IMAX theatre is that it should represent as closely as possible the normal real-life visual experience. Firstly, real life commonly involves a lot more looking down than looking up, so whereas in a conventional cinema most of the audience will look up at the screen, for a large part of the IMAX audience, the screen will be on a level with their eye-line, and they will be able to look down at the lower section of the screen. In fact the bottom of the screen will be partially obscured by the front rows of the audience. The idea is that there should not be a defined end, but it should appear that there is a little more that you could see if you could bend forward enough. Also, the steep rake of the auditorium allows the audience to be much closer to the screen than in a conventional cinema. The screen then fills almost the entire field of view and the IMAX spectacle can take over.

With all the elements in place, IMAX theatres have been a popular form of entertainment since their introduction at Expo '70 in Osaka, Japan, with Tiger Child showing to an audience in the Fuji pavilion who were carried through the theatre on a rotating platform. Subsequently the first proper IMAX theatre was built at the Ontario Place in Toronto, Canada, where the original IMAX projector from Expo '70 still operates. One of the most curious aspects of the medium, however, is the IMAX genre of film making, which is overwhelmingly documentary rather than dramatic. In fact, if anyone were to criticize IMAX presentation in general, it would be the films themselves that prompted the criticism. But we have to take the long view. The whole *raison d'être* of the IMAX experience is that it is a spectacle. You go into a special theatre to view a large format film. The opening credits roll on a screen that fills your entire field of vision. And when the helicopter clears the crest of the peak and you look down into the vast expanse of the valley below, your stomach raises to the top of your abdomen and its contents apply a gentle backward pressure on your

Figure 16.4 IMAX 3D camera. (©Imax Ltd 1999.)

oesophagus. How can mere wordplay between actors compete with that? Large format movies shot from the Space Shuttle are truly awe-inspiring. Breathtaking isn't the word when you can see the Earth from space as clearly as the astronauts see it themselves.

IMAX as a spectacle has a promising future since Imax Corporation have resisted any temptation to rest on their laurels and have developed IMAX Dome, a fisheye version of IMAX with a hemispherical dome screen. IMAX HD technology ups the frame rate to 48 fps, despite the technical difficulties outlined above, removing the flicker problem that large screens in particular suffer from. The IMAX 3D system employs two camera movements in one housing, a twin film projector, and either glasses with polarizing lenses, or a special virtual reality-like headset with LCD shutters controlled by an infrared beam in the theatre. As IMAX technology matures and more people have had the opportunity to experience the novelty of the spectacle, perhaps it will develop into a dramatic medium that will reach far beyond the capabilities of conventional cinema. There never will be an IMAX theatre in every town (although people once said that about the telephone), but when the opportunity to visit one presents itself, it is certainly not to be missed.

IMAX sound

A large-scale visual spectacle also demands large-scale sound. Sound in the cinema in 1970 was crude compared with today, and IMAX sound was certainly advanced for its day. The soundtrack of the film was not on the print. The format's running speed of 66 ips might present a certain wear factor on the soundtrack. Instead a 35 mm fullcoat 6-track sound film was originally synchronized to the projector. The six tracks provided left, centre, top centre, right, left surround and right surround channels. Interestingly, the top centre channel provides a vertical image lacking in conventional cinema sound. Of course, technological progress has moved beyond the capabilities of magnetic sound and new IMAX installations benefit from a CD sound system employing three synchronized discs. In its early days, IMAX sound would have been exceptional but now of course the Dolby, DTS and SDDS systems are strong competition.

Post-production

Interestingly, the full IMAX technology is used for shooting and projection, but not for rushes or editing. The inconvenience of providing rushes on a daily basis in 65 mm format, and then having IMAX projection (or at least viewing) equipment available really rules it out.

Fortunately, it is possible to print 15/70 images down to 35 mm so that rushes can be viewed on standard equipment. If editing is carried out using conventional film technique, then 35 mm work prints are employed. In all sections of film production, having rushes printed onto video is a possibility and offers significant convenience over traditional formats. But the difference between IMAX image quality and video rushes is so great that one has to wonder how worthwhile the exercise would be, although it certainly is done. Editing on nonlinear systems is also possible, and from a story-telling point of view there should be no drawback. Once again, since IMAX technology is so much more revealing, one has to wonder whether edit decisions made on a nonlinear system will extrapolate effectively to the large-scale medium. Changes could of course be made if necessary after the answer print stage.

CHAPTER 17

Telecine

The chances are that when you think of telecine, if you ever do, it is simply as a means of showing movies on television, but telecine is a creative tool in its own right. It can also be an extremely expensive tool – for the price of a top of the range telecine machine you could buy several digital video recorders, and they're not exactly inexpensive! As technology advances we are used to the new replacing the old, like digital video is rapidly replacing analogue, paralleling developments in the audio world. We may be tempted to think that film is an old-fashioned technology that will eventually be ousted by video. But film can still offer benefits unachievable in any other way.

Film feats

Telecine is very simple in essence and a broadcaster whose aim is to fill air time as cheaply as possible with movies bought at bulk rates is unlikely to be particularly interested in the finer points. But many people use telecine as an active and integral part of the production process. For this we need to look at film itself and understand why it is still widely recognized as an intrinsically superior medium to video.

If you have dabbled with photography or amateur video at all then you will know that one of the greatest problems is in getting an accurate picture when the contrast range of the subject is high – for instance when part of the scene is in the sun and part in the shade. If you live or work in a big city then every so often you might stumble upon a film or video shoot. I'm always amazed by two things, one being the size and power of the lights they use even outdoors in full daylight. The other is the size of the catering trailer. The function of the artificial illumination is not to make the bright parts of the scene brighter, it is to fill in the shadow areas and reduce the overall contrast ratio. On film, it is possible to squeeze a contrast ratio of around 3000:1 in the original scene down to 30:1 on the original camera negative, and since film is an almost linear medium this lower ratio may be expanded as necessary in the making of the print. In the cinema, allowing for flare in the projector lens and a small amount of ambient light, it is possible to experience a contrast ratio of 800:1. Compare this with the accepted value for video, even under the best conditions, of only 50:1. This small range reduces down even further in the average domestic situation.

We consider the small contrast ratio acceptable on the TV screen because we have no alternative, but it means that the right compromises have to be struck at the shooting stage, and the contrast of the original scene reduced with fill-in lighting if necessary. With film, in comparison, a much greater contrast range can be captured from the original scene, and corrections applied at leisure during the telecine process. Think of video's 50:1 range as a movable window onto the much greater range on film. If you need detail in the shadow areas, then move the window in one direction; if you need detail in the highlights, then move it in the other. This is a bit of a simplification, but I shall elaborate shortly.

Technology

Let us go right back to the beginning and look at the reason why telecine technology came into being in the first place. The final link in the chain in creating the first workable TV system was the camera. John Logie Baird's ultimately unsuccessful mechanical system lacked an adequate camera (later invented by Vladimir Zworykin and others) so he devised a system to substitute a film camera instead!

In his early experiments, Baird produced a video signal by shining a light through a disc punched with a spiral of holes, thus scanning the scene with the light creating a raster of lines (which was reportedly rather dazzling for the subject). The detector could be an ordinary photocell. This system was very simple and moderately effective. But when it came to the contest between rival television systems at the BBC's studios in London in the 1930s, to achieve better sensitivity to light Baird was using his new intermediate film process (Figure 17.1) which involved a fiendish machine shooting conventional photographic film, developing it in about a minute and then telecineing the result. History recalls that this machine took rather a lot of maintenance, and it is not hard to understand why the rival EMI Marconi all-electronic system won the battle for TV supremacy in the UK. There were early experimenters into mechanical television in the USA too, but Zworykin's Iconoscope allowed electronic television to flourish and prosper and its superiority was recognized almost from the beginning. Interestingly enough, Baird's company, the Baird Television Development Company, did not go broke as a result of this setback. In 1940, it was taken over by Gaumont British and renamed the Cinema Television Company. Now – further renamed – Cintel is considered one of the world leaders in telecine technology.

Some modern telecines still use Baird's telecine technique, known as 'flying spot' in which a very bright cathode ray tube, similar to the tube

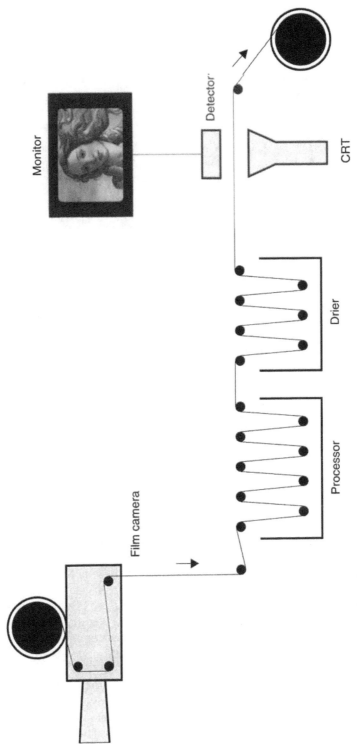

Figure 17.1 Baird's film/television hybrid system.

in a TV receiver, creates a source of light which scans the film. Other high-end telecines use CCD (charge-coupled device) detectors. Here I shall concentrate on flying spot telecine.

Cathode ray tube

If you look inside your desktop scanner you will see that the light source is a miniaturized fluorescent strip light. All the clever stuff is done in the CCD detector. Cintel, however, work on the principle of having an intelligent light source and simple (in comparison) image detection. The light source is in fact a bright, high precision cathode ray tube or CRT. The CRT shares many features in common with the CRT of a television receiver, but it displays no picture. There is only an electron beam which creates a uniform bright raster of lines on the face of the tube which is focused onto the film. The path of the electron beam is precisely controlled to create this line structure. On the other side of the film, the detector is a photoelectric cell. Earlier machines used photomultiplier tubes but now large area avalanche photodiodes are employed. The object is to convert the light passing through the film into an electrical signal with a good degree of linearity.

I said that the CRT bears no image. In fact considerable pains are taken to ensure that the light is uniformly bright. In principle the CRT is like any other television tube: an electron gun at one end firing at a phosphor-coated screen at the other. The tubes are made by Brimar, Cintel's sister company, using a dispenser cathode running at 30 kV for the source of the electron beam. The phosphor is carefully researched and formulated to achieve high light output, good resolution and long life. The phosphor is hand-coated onto the face of the tube for maximum consistency. A CRT for C-Reality costs of the order of $16 000, and for that you would expect to get around 5000 working hours.

In designing a telecine tube there are of course trade-offs to be balanced. More light is good, as it reduces noise in the detection and signal processing. It would be possible to get more light by throwing more electrons at the phosphor, but that would mean that the CRT would wear out more as the phosphor is burned away. It is another challenge for the designer. For this sort of application a very precise, tightly focused spot is required. Ideally the phosphor should glow brightly the instant that the electron beam hits it; and equally instantly emit no light when the electron beam moves away. In practice this perfection is not attainable, and there is some afterglow. Without electronic filtering in the subsequent signal processing, this would result in a softening of the image.

The real challenge for the CRT designer is to achieve absolute consistency of light output. Some inconsistency is inevitable: even with a very long tube, elementary geometry says that there is a path difference between the centre and edge of the CRT faceplate. The edges are further away from the gun, so get fractionally less stimulation, so emit fractionally less light. This variation across the face of the CRT is called shading, and is compensated electronically. In theory, inconsistencies or any phosphor granularity across the face of the tube would show up in the same place in every frame and would exhibit themselves as a fixed pattern. To avoid this, a system known as Scan Track is used where the scan patch (the bright area) is moved around the face of the tube, with appropriate compensation in the detector. This also reduces wear on the phosphor. When Cintel's earlier URSA model was launched, one of the revolutionary things about it was the shading map. At the start of each session it is put through an auto-alignment routine, in which the entire face of the CRT is scanned without film in the gate. The result is a detailed map of the tiny variations in light output (and any inconsistencies which may exist in the lenses in the gate). This is used to provide instant compensation by making appropriate adjustments in the detector gains. That calls for some very high-powered data processing in real time, and was recognized as a dramatic improvement at the time. C-Reality has the same system, but it is supplemented by real-time compensation. Nine fibre-optic light-pipes are arranged in a ring around the edge of the CRT faceplate. The light from the CRT, before it passes through the film, is taken to a second set of detectors which are identical to those used to capture the image. These give an instantaneous measurement of light output in red, green and blue. In effect, the image signal is now the differential between the light from the tube and the light from the film. This provides not just excellent shading but also long-term consistency so that a job can be put back on the telecine after weeks or months, and it will look exactly the same.

Scanning

As mentioned earlier, the required line structure is created by the CRT. It is easy to imagine how a still frame can be scanned in this way but of course the whole point of a telecine is to scan moving images. Some telecines are designed with intermittent motion: the film is advanced a frame, stopped in the gate, scanned, then advanced another frame, and so on. Very high precision scanners work this way, but they run very much more slowly than real time. There are telecines which can perform intermittent motion in real time, but it is thought by some to stress the film and is therefore not safe for an original camera negative. In C-Reality, the film moves at a constant speed through the gate and the scans are arranged to match that movement. In normal running on C-Reality the scan patch is larger than it is in still, because the light is playing catch-up with the film. In reverse running, the scan patch gets smaller. C-Reality currently offers running speeds forwards and backwards from 0 to 30 frames per second in 0.01 fps steps, with high speed options soon to be added. This part of the control system is far from trivial. In addition, it is quite common to scan just a part of the full frame. Other manufacturers use a downstream digital effects system to achieve this but the extra signal processing introduces some signal degradation. With the flying spot system, zooming and other effects such as rotation can be achieved in the scan, so that the signal is always at optimum quality. To zoom in means that the scan patch is reduced in size so that only the required part of the film frame is scanned.

Digital processing

Once the raw image has been captured then it requires further processing to reveal it in its true glory. At this stage we are talking about a 14 bit digital image in three channels: red, green and blue. Three significant items of terminology are lift, gain and gamma. Lift involves setting the black level of the video signal. The film will have many more shades of 'black' than mere video can register so it is necessary to mark a cut-off point in brightness below which all dark and very dark greys will translate as black. Once this fixed point is set, gain is applied effectively to set the point in brightness above which all very light greys will appear as peak white in the video. This is all very logical, but perhaps the most interesting is the gamma control. When the ultimate boundaries of black and white have been set, the colourist will undoubtedly find that some tones which are clearly differentiated on the film are exactly the same on the video due to its lower resolution of contrast, so he or she may – within the limits of the black level and white level that have been set – increase the contrast of the mid-tones by applying an S-shaped contrast curve. Further than that, it is possible to tweak the contrast in the highlights or 'lowlights' to keep as much detail as possible in the most important areas of the picture. If this seems complex, remember that the film has almost the entire contrast range of the original scene and there is almost certainly more time available in the telecine suite to squash it down onto video. In C-Reality all the contrast and colour correction is done digitally and to make this possible the raw digital signal has 14 bit resolution on each of the three colour channels to capture the full contrast range of the film. The more bits you have, the more flexibility there is for further processing and effects. Matting in particular works better on a higher resolution signal than if it is performed on an 8 bit or 10 bit signal of the same standard as the final output on digital video tape, particularly if the colour bandwidth has been reduced.

Early colour telecines offered only processing of the red, green and blue channels. URSA, Cintel's previous top of the range model, had secondary correction for cyan, magenta and yellow. C-Reality goes one stage further with vector processing, which means that any colour or any group of colours can be isolated and manipulated. A good example would be a

Figure 17.2 The Cintel C-Reality telecine.

commercial with a pack shot featuring a distinctive shade of red. Suppose a woman handling the pack wore a red dress and had red lipstick, but they didn't quite match the pack. The colourist would be able to isolate the colours of the dress and lipstick and change them to match the pack exactly. This isn't new – it has been offered in colour correctors from Da Vinci and Pandora – but this is the first time it has been offered inside the telecine where the image is still at 14 bits per colour resolution. As an example of the consideration given to colour grading, for a typical fifty-minute TV programme shot on film, something like a day will have been spent in telecine grading each scene, and each shot within each scene.

High resolution

Telecine for standard definition television is one thing. In the past, telecine for work at higher resolutions was quite another, but now the entire spectrum is converging and a high-end telecine such as C-Reality can cover the full range. When digital film effects first started to gather momentum, the manufacturers did their calculations and decided that the ideal resolution would be 4096 pixels across the width of a 35 mm frame (remember that this is around half the size of the frame in your 35 mm stills camera). Unfortunately, this produces a file size of around 75 Megabytes per frame, which is simply too much data to deal with sensibly. So a compromise of 2048 pixels has been adopted as something of a standard, known just as '2k resolution'. C-Reality also features 'resolution independence', which has to do with the method of scanning. A CCD telecine has a fixed 2k optical resolution, which means there are approximately 2000 individual sensors in a line that scans the height dimension of the frame. To produce a standard definition television image this resolution is digitally reduced to one-quarter. This is not a problem since digital processing can be very clever these days. Neither is it a problem to zoom into the film image at standard definition because there is resolution to spare. C-Reality can, however, zoom into a small part of the image at full 2k resolution. Panning and zooming are in fact powerful techniques which are commonly used in telecine, so this is an obvious advantage.

The creativity of telecine is demonstrated most in digital effects for feature films and in commercials. Particularly in TV commercials, a lot of the effects that you might imagine are done with clever video processing are in fact created, or at least facilitated, in telecine. Positioning the image correctly, aligning images for morphs, etc. and particularly zooming into a smaller part of the complete film frame are far better done in the high resolution telecine environment than left until later. The art of the telecine colourist is deep and precise, and often insufficiently appreciated. My prediction is that despite high definition television, and other advances in electronic motion picture imaging, film has a long future in front of it, and telecine will remain a vibrant and cutting edge technology.

CHAPTER 18

Pulldown

'Up a bit . . . down a bit . . . a bit more . . . OK that's it, lock it there.' The sluice gate operator of the Hoover Dam followed his instructions precisely and the 60 Hz mains frequency of the USA was established. A similar scenario played out in Europe and a 50 Hz mains frequency was chosen. Half the world followed the lead of the USA, the other half followed the European standard, and the eventual market for video standards converters was created. A canny investor might have spotted the potential!

A while later in the USA, a member of the National Television Standards Committee commented, while viewing an early experimental colour image: 'What's causing that diagonal interference pattern?' A few quick tweaks of a reference oscillator and the frame rate was reduced from 30 fps down to 29.97 fps and, little did they realize at the time, the world was condemned to a future of increasing incompatibility, and general fear, uncertainty and doubt. Before the standards for television were set in stone, why on earth didn't anyone consider the universal frame rate of 24 fps for sound motion pictures, recognized the world over? Perhaps the technical difficulties with early systems seemed more important at the time than potential problems that might occur in the future, but no one could have guessed the scale of the problem, the amount of effort put into designing solutions (often only partial solutions), and the time wasted correcting mistakes that would never have occurred if there was a single unified frame rate for film and video the world over.

Since the USA has been blessed with a non-integer frame rate, pulldown is a bigger issue there than it is in Europe. As we shall see, however, PAL users do not escape entirely, but the need for pulldown in the first place arose even before NTSC when the US video frame rate was a nice round 30 fps. The question was: 'How do you show a film running at 24 fps on TV running at 30 fps?' The easiest answer might be to speed up the film to 30 fps and show it with a 1:1 relationship between film and video. Unfortunately, everything would be so fast it would look like Mack Sennett and the Keystone Kops, so that idea was a non-starter. Fortunately there is another way of doing it which is still relatively simple. Film, as you know, runs at 24 frames per second, which is sufficient – just – to give the illusion of smooth motion. It isn't sufficient, however, to prevent the projected image flickering, so a rotating shutter in the film projector splits each frame into two, giving a flicker rate of 48 Hz. Once again this is just sufficient. In television and video, even the faster US frame rate of 30 fps isn't enough to avoid flicker so a system of interlacing is used where first the screen is filled with half the available number of lines, leaving gaps in between, then the gaps are filled in by the remaining lines to complete the frame. Two half-resolution images are therefore displayed in every frame and, from a 30 Hz frame rate, a 60 Hz flicker rate is created. Each half-frame is known as a field. The PAL system is the same except the frame rate is 25 Hz and the field rate 50 Hz.

To transfer 24 frames of film to 60 fields of video is now a more manageable proposition, even though 24 still doesn't divide evenly into 60. All that is necessary is to repeat frames at regular intervals so that one frame of film on average covers 2.5 fields of video. This is done by matching one frame of film to two fields, then by stretching the next frame over three fields, the next frame covers two fields, and the next frame, once again, covers three fields. Hence we have the sequence known as 2:3 pulldown (as shown in Figure 18.1). You may have heard

Film

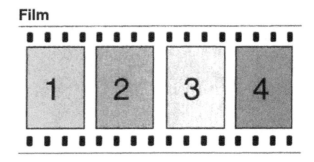

NTSC video

Odd fields:

Even fields:

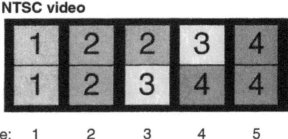

Video frame: 1   2   3   4   5

Figure 18.1 2:3 pulldown.

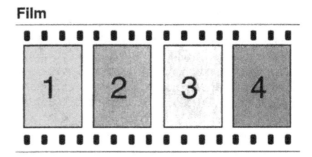

Figure 18.2 3:2 pulldown.

this expressed as 3:2 pulldown. The two expressions mean slightly different things, and even when people talk about 3:2 pulldown (see Figure 18.2) they commonly mean 2:3. Fortunately the people in the Telecine suite know all about this and can make the necessary adjustments. Here we shall be correct and stick to 2:3 when that is what I mean.

Looking at this in theory, it is hard to imagine how it will work out in practice. It sounds a bit start-stop and potentially visually obtrusive. However, it does work out in practice, as is demonstrated by the fact that, whenever material sourced on film is shown on TV in NTSC territories, it all looks fine and no-one calls the network to complain. There are worse problems to contend with in PAL territory!

29.97

It's that number again. So far, what I have explained relates to the late lamented 30 Hz frame rate. Well, that's gone and there is no prospect of it making a comeback. Television and video in NTSC territories run at 29.97 frames per second, and that will be that for the foreseeable future. So how does pulldown work now? The answer is simple in essence, but complex in its ramifications. To transfer 24 fps film to 29.97 fps video, all that is necessary is to run the film slightly slower in telecine – at 23.976 frames per second in fact. The 2:3 pulldown sequence works as explained and the result is a transfer that works fine and looks great, with the only proviso that the video runs 0.1 per cent slower than it actually should. 0.1 per cent is neither here nor there in human terms. People who can tell the difference are exceptional in their sensory abilities. It is, in theory, possible to detect a 0.1 per cent difference in pitch at mid-frequencies, which equates to the difference between 1000 Hz and 1001 Hz, but there are few who would make an issue of it.

A difference of 0.1 per cent might not amount to very much, as long as both the video and audio run at the same speed. But, if the video is slowed down by 0.1 per cent, and the audio stays the same, then drift is going to occur which is pretty soon going to become exceedingly noticeable. There are two strands to this problem: analogue and digital. Analogue audio is easier because it does not, in itself, have any kind of clock or frame rate. Audio for film in the USA is recorded against a 30 Hz timecode frame rate (not the 24 fps that you might have expected) or, on a Nagra portable recorder (one of the hundreds of examples that are still in regular daily use at the highest level of film making), there might be 60 Hz Neopilot rather than SMPTE timecode. (Neopilot is an updated version of Pilotone where a pulse, generated by the camera or crystal oscillator, is recorded on the tape and is later used as a reference so the audio can be transferred to the editing medium – mag film or hard disk – in sync with the pictures. A clapperboard gives the positional reference.) If the sync reference used is Neopilot then, during transfer, the Nagra must be locked to an external 59.94 Hz crystal to bring it down to the correct speed. If 30 fps SMPTE timecode was used on the shoot then, during transfer, the machine must be synchronized to 29.97 fps code. With digital audio the situation is similar: because the film is slowed down by 0.1 per cent the audio must be slowed down by exactly the same amount. The only problem we get now is that there are sample rates involved! Going back to square one but in the digital domain . . .

Suppose a film is shot at 24 fps and audio is recorded at a sample rate of 44.1 kHz. When the film is transferred to video at 29.97 fps, undergoing a 2:3 pulldown and in addition being slowed down to 99.9 per cent of its original speed, and when the audio is slowed down by the same amount, the sampling rate becomes 44.056 kHz. If a 48 kHz sampling rate had originally been used then the new rate will be 47.952 kHz. Confusingly, where pulldown in video means converting from 24 fps film to 30 fps or 29.97 fps video, the word 'pulldown' in audio is usually taken to mean the slowing down from 44.1 kHz to 44.056 kHz or 48 kHz to 47.952 kHz. The easy solution to this is, firstly, to use equipment that will allow the pulled-down sampling rates – this equipment isn't common, but it is certainly available – and, secondly, to make the transfer to DVTR via analogue cables! While the idea of recording digitally on location and editing digitally all the way through the project is very appealing, having to drop into the analogue domain for the telecine transfer doesn't seem quite the right thing to do.

But who is going to know? Who in the listening public can actually tell whether the location audio was recorded on a Nagra or on DAT? Of course, where there is a problem, a true engineer will seek a proper solution and not just a bodge. There are ways and means of transferring digitally and ending up with everything in sync at the right sample rate. The way to do it is to record digital audio on location with a 0.1 per cent pullup, so the sampling rate is higher than normal. 48 kHz becomes 48.048 kHz. It's a good job that most people are not worried about a 0.1 per cent pitch difference because the opportunities for error here are manifold. When the film is transferred at 29.97 fps, the audio is simply slowed down to 48 kHz which can be transferred digitally to a DVTR. Easy(!)

Shooting to playback, as in a music video shot on film, means more pulldown adjustments. The record company supplies a DAT with 29.97 fps timecode, playback takes place at the pulled up 48.048 kHz, and the action in front of the camera takes place 0.1 per cent faster than it ought to have done. In transfer the film is slowed down from 24 fps to 23.976 fps, and the audio is transferred digitally at its original 29.97 fps, 48 kHz rates.

PAL pulldown

The problems, though, are not restricted to NTSC. Far from it. The relationship between the 24 fps film frame rate and 25 fps PAL may be very much closer than in the NTSC system, but that makes it very much more complex. The standard solution is to pretend it doesn't exist. Film shot at 24 fps is telecined to 25 fps video and shown on PAL (and SECAM) televisions at that rate. In other words, anything shot on film and shown on PAL TV runs approximately 4 per cent faster. Earlier on I said that a 0.1 per cent change didn't amount to anything. Well, a 4 per cent change certainly does. Fortunately (for the broadcasters), most viewers are blissfully unaware that every film they see is speeded up, but it certainly does affect the experience, and not for the better. Firstly, a film that was intended to last for 100 minutes will only last 96 minutes. Then the audio will be significantly higher in pitch. Try a 4 per cent increase on your pitch changer and see what effect it has on the human voice. A noticeable effect, you will agree. Music too – anyone who has tried to strum their guitar to the title song from High Noon will have realized that it's impossible without retuning the guitar, since a 4 per cent pitch shift is roughly two-thirds of a semitone.

The people who notice this PAL speed-up most of all are film editors – people for whom timing is all-important. On slow-moving scenes it is not such a significant difference, apart from the factors I have mentioned above but, in action scenes, it is crucial. To create a fast-paced action sequence an editor will trim each element down to the shortest it can possibly be. They will try it a frame shorter – physically or mentally – decide that it doesn't work, and then set it back to exactly the right number of frames. When it is played on PAL TV it looks like the aforementioned Keystone Kops! We don't notice because we are used to it, even if we watch a movie on TV that we recently saw in the cinema, but it certainly isn't correct and technically-aware visitors to Europe will see the problem. The irony is that even movies made in Europe suffer this problem when shown on TV, and they are shown at the correct speed (less 0.1 per cent) in the USA.

The process of making production video masters involves colour correction and re-composing for the different aspect ratio of TV. Producers and directors want to be involved in this, simply because a movie is a work of art and they want home viewers to have as good an experience as they can possibly have within the limitations of video and TV. For a movie made in the USA, the NTSC master will be made first and the colour correction and other information simply transferred over to the PAL master when it is made (hopefully as a separate process and not as a standards conversion). By this time the director and/or producer will be yawning and probably regard the rest as simply a technical exercise, and hence never realize how different their film can look (they might also consider that the phosphors in a PAL receiver have different colour values to those in an NTSC set). Transferring audio at 4 per cent faster than normal speed is not a great problem in the analogue domain, considering that all hope of keeping the pitch correct has been abandoned. Transferring digitally is a more interesting problem since the number of digital recorders that can synchronize, not just varispeed, at this higher rate is limited.

There are, strangely enough, some advantages to the PAL speed-up. Some people actually think it is more appropriate for action to be a little faster on the small screen. Compared with the alternatives there are no visible artefacts generated since the frames correspond 1:1 between film and video. Transfer times are shorter, stock costs are reduced and, if it was not for restrictions on broadcasters, there would be more time for revenue-generating advertisements!

I think, since it is established that there is a problem with showing films 4 per cent too fast, that surely there must be a solution on the horizon. In fact, pulldown can be applied to PAL just as it can to NTSC except that the figures are different. PAL pulldown (Figure 18.3) is known as 24&1, meaning that frame 12 of a sequence is extended to cover three fields rather than two, and the same is done to frame 24, with the result that 24 frames of film are stretched over 25 frames of video and the running time is correct. One other, more straightforward and accurate solution is simply to run the film camera at 25 fps so that the film can be telecined

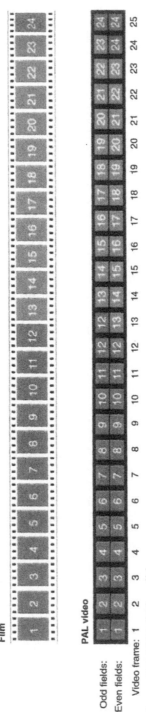

Figure 18.3 24&1 PAL pulldown.

frame-for-frame. This is fine, but what are North Americans going to make of it? Rather than engineer a pulldown to convert 25 frames to 30 they would probably rather standards convert the video.

So what should you do?

I could advise to avoid pulldown situations at all costs, but life is not like that. The best advice is to be aware of pulldown and its implications but, before that, there is one thing that is so important that it should be made law, that is, always to refer to timecode by its real frame rate. A large proportion of timecode misunderstandings arise from people saying '30 drop' or similar when they mean 29.97 fps drop frame timecode. That difference of 0.1 per cent might be small but, in sync terms, it certainly does add up. You should also remember that there is such a thing as true 30 fps drop frame code and the distinction is vital. Beyond that it is important to know at what speed the camera is running. A film camera anywhere in the world almost certainly will be running at 24 fps, but it just might not be. The problems that are intrinsic in mixing film and video have been solved by many people in many ways and, unfortunately, the variety of solutions has thrown up more problems that have to be dealt with one way or the other. If film is being shot for transfer to NTSC, and if the camera is running at 24 fps, then analogue audio can be recorded at normal speed and slowed down unnoticeably during transfer to video. Digital audio can be recorded at one of the normal sampling rates, slowed down by 0.1 per cent during transfer, and an analogue signal taken to the DVTR. If a digital transfer is deemed necessary then a sample rate converter will have to be used, or the recorder pulled up to 48.048 kHz during shooting. In PAL, the likelihood is that either 24 fps film will be transferred frame-for-frame to 25 fps video, resulting in a speed and pitch increase, or the camera will run at 25 fps. If 24&1 pulldown is used then the speed correction will be small and hopefully within the sync range of the recorder but, once again, analogue transfer or a sample rate converter would be necessary.

A new frame rate?

This could be the last thing we need since a new frame rate will surely lead to a new timecode frame rate. In the USA, film cameras run at 24 fps and TV and video run at 29.97 fps. This discrepancy can mostly be dealt with by means of pulldown. There are, however, reasons for shooting at other than the normal film frame rate. One is when there are TV monitors in shot, which is not an uncommon occurrence. When a TV monitor is in shot, and the film is transferred to video through the normal 2:3 pulldown there will be a strobing effect where the film and video frame rates clash. The solution is to slow down the film during shooting to 23.98 fps (as it is usually expressed – it is rounded up from 23.976 fps) and, like magic, the figures will work and the problem disappears. Another reason for shooting at 23.98 is to ensure correlation between the film and video-assist images. 23.98 is, so I understand, set to become significant, if not commonplace.

CHAPTER 19

Lighting technology

Film and video lighting

In the early days of motion pictures, lighting was provided cheaply, but maybe not so conveniently, by the sun. Hollywood sunlight has the advantages of being bright and readily available, which suited motion picture pioneers, but also has the distinct disadvantage that it is not very controllable. The sun also has an annoying tendency to move! Early film producers sometimes went to the trouble of constructing rotating sets so that they could get as much consistency as possible during the working day. In parts of the world with less consistent weather conditions, the quality of natural light varies enormously. When the sun is shining and there are no clouds in the sky, the light source is effectively a point source which will cast harsh shadows, as shown in Figure 19.1(a). When the sky is overcast, then the source of light effectively becomes larger and the edges of shadows are softened, as in Figure 19.1(b). A light source that emits light over a wide area is often described as a soft light. When there is direct sunlight, but there are also some fluffy white clouds in the sky, then some of the light will bounce off the clouds making the shadows a little less dark than they would otherwise be. In fact, although there is no problem at all about sunlight being bright enough for filming, it can create shadows which are so dark in comparison to the illuminated areas that no film or video camera can capture the range of contrast produced.

The answer is to use artificial light to 'fill in' the shadows – not to such an extent that they are not shadows any more, but just enough to balance the light from the sun. The sun, however, is so bright that exceedingly powerful lights have to be used for fill-in. It seems like a paradox, but the brighter the sunlight, the more artificial light that has to be used on an outdoor shoot. In the early days of motion pictures, carbon arc lights were used which could indeed give the sun a run for its money. A carbon arc light uses two carbon electrodes which are separated by a short distance. The electric current will jump between the electrodes forming a very bright arc and a considerable amount of smoke. Although the quality of arc light is sometimes desirable in its own right, the disadvantage is that

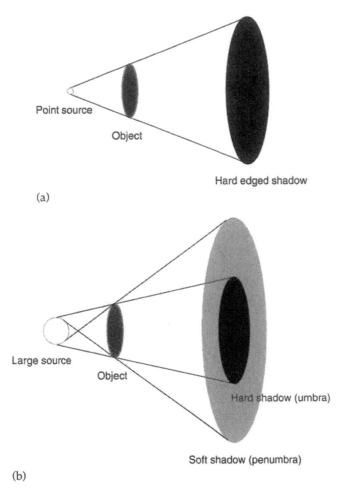

Figure 19.1 (a) Hard-edged shadow; (b) soft-edged shadow.

each light needs an operator to adjust manually the electrodes as they burn away. I am sure that modern technology could provide a solution, but there are now alternatives that are not so much trouble and don't produce any atmospheric pollution, on site at least. The biggest of the arc lights, by the way, are called 'brutes' and run at a current of 225 amperes and 72 volts d.c. This equates to just over 16 000 watts of power. Impressive?

Lamps glow, bulbs grow

Carbon arcs have had their day it seems. Parallel with their development was the improvement of the incandescent tungsten filament lamp. I've got to be careful with my terminology here. 'Light' is the generic term for the whole piece of equipment, including the lamp and the housing, except in the theatre where a light is a 'lantern'. To confuse matters further, manufacturers – but hardly anyone else it seems – tend to refer to lights as 'luminaires'. Americans may also call a light a 'fixture', even if it is on a movable stand. Even the word 'bulb' is not in universal use in the professional's vocabulary, many insisting on 'lamp', or sometimes in the USA, 'globe'. You may come across the word 'bubble', too.

An incandescent lamp is simply a heated filament enclosed by an inert gas in a glass envelope. As a filament is made hotter, obviously its light output increases but also its colour temperature increases too. Colour temperature is a measure of the bias of the colour output of the filament towards different parts of the visible spectrum. At lowish temperatures, the filament will glow dull red, then as it becomes hotter the balance will sway towards the blue end of the spectrum. In theory the actual temperature of the filament in Kelvin (a change of one Kelvin is the same as a change of one degree on the Celsius scale, but absolute zero is 0 K rather than –273 °C) relates to the colour balance, so it is very convenient to describe a light source as having a colour temperature of, say, 5400 K, which as it happens is the colour temperature of natural daylight in Washington DC at noon on the average day, according to the *American Cinematographer Manual*. Tungsten lamps are commonly made with a colour temperature of 3200 K, which is rather different to the colour balance of natural daylight under most conditions. Of course this is a nominal figure that depends on the output of the lamp.

The output of a lamp is measured in lumens and in the case of incandescent lamps this depends almost entirely on the temperature of the filament, which in turn depends on the voltage supplied to it. Light output and colour temperature are, however, linked. A change in voltage of around 1% can result in a colour temperature shift of 10 K; a drop to half the rated output can change the colour temperature from 3200 K to under 3000 K, which is on the borderline of acceptability in film and video. In some cases, however, such as illuminating the human face, apparently a colour temperature shift down to 2750 K is acceptable, which would correspond to around 25% of a lamp's rated output. What this means in practical terms is that in the theatre, where the scene is viewed directly by the human eye, the lamps can be dimmed all the way from full output down to zero as necessary, as long as the lighting designer thinks the results look good. When the scene is to be captured on film then dimmers can be used, but only up to a point before the results would be unacceptable. In fact, the easiest way to change the brightness of a light is simply to move it closer or further away, and this is often the technique employed. Moving a light – sorry, lantern – in the theatre is not usually as practical as dimming it because of the limitations of the building and the requirements of the set. Also, in film and video, shooting may be done outside the studio and artificial light has to be balanced with daylight. This is done by filtering the light to shift the colour temperature upwards, at the expense of losing a proportion of the light's output. It is also possible in certain cases to make changes to natural daylight. Indoors, a filter can be placed over a window to convert daylight to a tungsten colour balance. Or if you wish that the sun was not quite so bright, you can 'turn it down' a bit by shading the scene with 'scrim', which is a fine mesh that blocks out some of the light.

From the tungsten filament lamp, the tungsten halogen lamp has been developed. This uses a similar filament, but the lamp has a thick, hard glass or quartz envelope and is filled with high pressure halogen gas, usually bromine. The principal advantage is that the tungsten that evaporates from the filament is redeposited back onto the filament rather than onto the inside of the envelope, hence there is no blackening during use. Tungsten halogen lamps therefore maintain their output and colour temperature throughout their life, and in fact have a longer life than conventional lamps. For the halogen cycle to work and redeposit tungsten on the filament, the temperature of the envelope has to be high, therefore these lamps are made quite small compared with their conventional equivalents.

HMI

Although carbon arcs have their limitations, the principle is sound and they have developed into the 'hydragyrum medium arc length iodide' or HMI lamp. HMI lamps were developed by Osram for the Munich Olympics in 1972 and have become enormously popular since then. Whereas a carbon arc is struck in air, the arc in an HMI lamp is struck in mercury vapour with additives to improve the colour balance. The colour temperature of lamps of this type is usually rated at 5600 K, but with a tolerance of plus or minus 400 K, which can be significant. Particularly if two HMI lights are to be used as the key light of a scene, careful colour temperature readings will be taken and filters used as appropriate. HMI lamps require a start-up period before they reach normal operation, and a higher voltage is necessary to strike the arc than is needed to maintain it. Typical ignition voltages range from around 20 V to 40 V, while operating voltages are 8 V to 20 V. HMI lamps are available from a modest 200 W all the way up to a potentially dangerous 18 kW. Eighteen kilowatts is the same as six three-bar electric fires, so you probably wouldn't want to stand in front of a light like this for too long. More importantly, however, HMI lamps emit large quantities of ultraviolet radiation which could easily damage skin, therefore luminaires are designed so that there is no possibility of opening the casing while the lamp is active. The lens or cover glass will screen out ultraviolet from the normal output of the light. The lifespan of lamps such as these is governed by the number of starts and hours of operation, and the colour temperature will drop about

1 K for every hour of use. The typical rated life is 500 to 750 hours, and at the high power end we could be talking about £1000 for a replacement!

Fluorescent lamps in the past have not been ideal for photography, video or film. The problem is that they are not incandescent lamps but 'discharge' lamps. This means that they do not have a continuous spectrum, and the term 'colour temperature' does not strictly apply. Although the light from fluorescent lamps may appear acceptable to the eye, careful filtering is necessary to achieve a reasonable result on film. Having said that, long thin lamps do give a very good soft light, and fluorescent lamps have been developed that are suitable for cinematography, but working out the filtration necessary to balance fluorescent lamps correctly with daylight and incandescent lamps is tricky.

Broads, Blondes and Redheads

Any specialized discipline develops its own peculiar terminology, and the terms that have come into popular use in film lighting are just a little odd. I'm sure you have heard of these three, so a few words of explanation may come in handy. These are all products of Strand Lighting, one of the top manufacturers of luminaires, and these products have become so successful that the product names have become generic terms for the same type of equipment from whatever manufacturer. A Broad is simply a 1.2 kW floodlight providing an even light output over a wide angle. The original Redhead, produced in the 1960s, was an 800 W floodlight that happened to have a housing made of red plastic. The Blonde was a 2 kW floodlight with a metal housing painted yellow. Pretty obvious really!

Compared with high-tech performance lighting, film and video luminaires are usually fairly simple devices. In luminaires that are designed to give a soft light, the object is to spread the light from the lamp over as wide an area as possible, so the design of the reflector is crucial. If a soft light had too bright a hot spot in the centre of its coverage then this would be a drawback. The larger the effective area of the light source, the softer the illumination and the less noticeable the shadows.

Lights other than soft lights come in several varieties, of which I shall describe the three most common. The simplest is the open face spot/floodlight which has no lens and is shown in Figure 19.2. The width of the beam is controlled by moving the lamp with respect to the parabolic reflector. The edge of the beam is neither well defined nor well controlled but efficiency is high since there is nothing to block the path of the light. Often a 'barn door' will be used to control the beam, but since light comes from the lamp and from the reflector, a double shadow from the edge of the doors may be noticeable. Figure 19.3 shows a 'fresnel' spotlight. A fresnel lens, first developed by Augustin

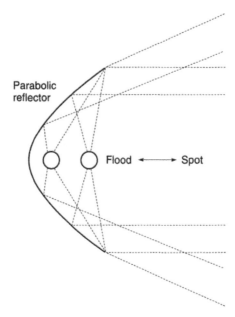

Figure 19.2 Open face spot/floodlight.

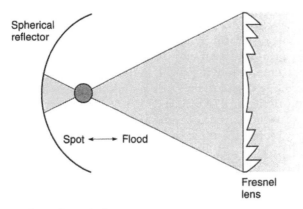

Figure 19.3 Fresnel spotlight.

Fresnel (1788–1827), is a plano-convex lens built up in segments so a great volume and weight of glass is not required. A fresnel lens would not be of any use for photography, but for focusing light it is ideal. Here the lamp and reflector are fixed with respect to each other, and the distance between the lamp/reflector assembly and the lens may be changed. When a very small tungsten halogen lamp is used it is possible to move the assembly very close to the lens, resulting in a very wide light distribution. Another advantage the fresnel system has over an open face luminaire is that barn doors can give a very sharp cut-off in the wide flood position.

Used without barn doors, a fresnel spot produces a soft-edged beam, which is good because one light source will merge unobtrusively with another. For some effects, however, a luminaire that can provide a hard-edged beam is desirable. A 'profile' spot uses a system of reflector and lenses very much like a projector, and indeed it is very often used to project a pattern, such as simulating light from a sunlit window falling on a wall. In this case a simple cut-out of the window shape is made from metal foil and is placed in the gate of the profile spot. The projected image can be sharply focused or slightly defocused as required.

You have probably already noticed that a surprising number of lighting people can turn their hand to sound when the occasion demands it. Now maybe you can start thinking about getting your own back! But lighting is a complex and developing technology. You may see thirty-year-old luminaires in use far more often than we use thirty-year-old microphones but that doesn't mean that the state of the art doesn't change. There is a tremendous amount of development going on in lighting, and the parallels between some of the technologies of lighting and sound are fascinating.

Moving light, changing colour – performance lighting

September 26, 1981 was a great day for performance lighting. Forty luminaires of a revolutionary new design were used on the first night in Barcelona of Genesis' Abacab world tour. Before this date, lighting was largely static in terms of colour and direction, but on this glorious night we entered the era of moving light and changing colour with the introduction of the first prototype VARI*LITE luminaires. It is true to say that the lighting designer's world, and that of the concert and theatre-going public, would never be the same again.

When I first explored the topic of film lighting I had been amazed to discover how simple it was – simple in technology, not necessarily in artistry of course. If a light is too bright on a film set, you simply move it further away, or place a 'scrim' in front which partially blocks the light. You can't use a dimmer to reduce the quantity of light because that would lower the colour temperature of the lamp filament, severely affecting the colour balance of the image recorded on the film. Video cameras can be balanced to colour temperatures other than daylight or normal tungsten illumination, but if two lights were used, one at full intensity and the other dimmed, then it would be impossible to balance for the different colour temperatures simultaneously.

Theatre and concert lighting has no such constraints because every-thing is done to look good to the human eye, and the eye is very much more tolerant of colour temperature than film or video cameras. Performance-based television lighting (for a popular music show for instance) may not have quite such freedom but the principle is similar.

Consequently, where film lighting consists largely of the well established technology of Brutes, Blondes and Redheads (not forgetting the Inky Dinky of course!), the technology of performance lighting is at a cutting edge every bit as sharp as that of digital audio, perhaps more so.

Going back once again to the early 1980s, Rusty Brutsché and Jack Maxon, two of the directors of the sound and lighting company Showco, were having lunch and discussing the feasibility of incorporating the then newly available dichroic filters into a luminaire and perhaps including a motor to change the colour of the light remotely. Suddenly Maxon said, 'Two more motors and the light moves.' It was a simple idea, like all the best ideas of course, but the difference between rigging and focusing a conventional lantern in a fixed position with a fixed beam, and being able to control the beam of a motorized luminaire remotely has made an immense difference to the possibilities of lighting design. Brutsché flew to England and demonstrated the first prototype to Genesis in their rehearsal studio in a centuries-old barn. This demonstration sparked a long-standing relationship with the band that still continues. Genesis' manager Tony Smith thought up the VARI*LITE name. By the way, if you are at any stage confused about the difference between 'VARI*LITE' and 'Vari-Lite', the former – apparently – is the product and the latter is the company.

The two most fundamental parameters of sound are level and frequency. We may change the overall level of a signal with the fader, or change the level at certain frequencies with EQ. Effects units allow us to delay a signal and play around with the timing too. Light has a number of parameters which are under the control of the lighting designer and operator during both the planning and performance of a show. The most basic of these are intensity (analogous to audio level) and colour (similar to audio frequency but just a little more complex). As I said earlier, the intensity of a light can be controlled by moving the luminaire or by dimming it. Moving a light closer or further away may be suitable for film and video shoots but it is obviously entirely impractical for live performance. Dimming the lamp by reducing the voltage and current is commonplace but not universal. Dimming by voltage change, as I said, lowers the colour temperature of the light. To compare this with audio, imagine that you have a music signal going through a graphic equalizer which is set a smooth curve peaking at 10 kHz. You would hear all the frequencies, but the upper frequencies would be emphasized. That would correspond to a high colour temperature where the light is bluish but still contains all the colours of the visible spectrum. Change the equalizer to a similar smooth curve with a peak at 200 Hz and you have the audio equivalent of a low colour temperature where everything has a reddish tinge, like at sunset. In theatrical performance, changes in colour temperature may not be entirely unnoticeable, but they don't stop anyone using dimmers as a means of controlling intensity. On the other hand dimming has its drawbacks. Not all lamps can be dimmed satisfactorily. Arc lamps, for example, will only go down to about 30% of their full output, which is not a huge change visually. Also, tungsten filament lamps take a little time to cool down and therefore cannot be dimmed instantaneously. The alternative is to incorporate a mechanical shutter which is very much faster. Obviously such a shutter has to be motorized otherwise the intensity of the light could not be controlled remotely.

The colour of light corresponds to its frequency, but that is a very simplistic viewpoint which only works if you are dealing with pure spectral colours. In the real world – if the theatrical world can be considered real – colour consists of hue and saturation. Hue is the spectral colour, and saturation refers to the amount of white light mixed in with it. A higher degree of saturation corresponds to a purer colour. But the eye does not detect hue directly, it has three sets of colour receptors which are sensitive to what we call the primary colours, red, green and blue. Spectral colours other than these will stimulate two sets of receptors, and the extent to which each is stimulated tells us the colour's position in the spectrum. For example, a sodium street lamp produces spectral yellow (actually two hues of yellow but very close in frequency) which stimulates the red and green receptors. Consequently, spectral yellow can be simulated by mixing red and green light in the correct proportions. Conventional lighting technology uses 'gel' filters where a sheet of coloured plastic is inserted in a frame in front of the lens of each lantern. A wide range of colours is available from a number of manufacturers, notably Lee Filters, but such filters have two main drawbacks. Firstly, the heat output of theatre lanterns is considerable and the colour of the gel changes over a fairly short period of time. Secondly, the gel may get so hot that it creates smoke! Though you might think the days of gel filters should be numbered because of such problems, they do have the advantage of simplicity, compatibility with probably hundreds of thousands of lanterns worldwide each of which has a potential life span of decades, and also – most significantly – many lighting designers still visualize colour in terms of gel filters and might request, say, a Lee 201 rather than try to describe the colour it represents.

Technology is irresistible, however, and modern luminaires such as VARI*LITEs use dichroic filters, which consist of very thin layers of metallic oxide coated by evaporation onto glass. Dichroic coatings, because of interference effects, transmit certain colours according to the metal oxides used and reflect others, and a single saturated wavelength can be achieved. Dichroic filters have the advantages of consistency and heat resistance, and therefore are very suitable for incorporation into modern luminaires. Since the colour is produced by interference rather than absorption, the colour does not fade and dichroic filters do not need to be replaced as gels do. VARI*LITEs use dichroic filters in two ways. One method, which is found in the VL2 luminaire, is to use two colour wheels, each of which has sixteen positions providing for 256 different combinations. One wheel consists mostly of blues and the other ambers and in combination they can produce a wide, if not infinite, range of hues and saturations. The other technique, as found in the VL4 and VL6 luminaires is Vari-Lite's proprietary Dichro*Tune system with three dichroic filters, each of which can be twisted in the beam which changes the saturation of the colour. The three filters are coloured magenta, cyan and amber, which correspond closely to the three secondary colours. For example, you can mix magenta and amber to get red. The reason why the filters are not precisely the secondary colours is that it is impossible to simulate every spectral colour by combining either primaries or secondaries. In television, for instance, combining light from phosphors which are close to the red, green and blue primaries can produce most of the colours we see in real life but not all, and even if exact primary colours were used there is a range of purple shades that cannot be reproduced. Vari-Lite's modified secondaries have be chosen so that the purples can go all the way from a light lavender to a rich blue and warm red shades are handled well too. The trade-off is that the range of greens that is available is compromised to some extent.

Intensity and colour are parameters available in conventional lanterns, but not under automated control. The same applies to pan and tilt. Obviously any conventional lantern can be precisely directed according to the wishes of the lighting designer. But once it is locked off then the direction of the beam is almost certainly going to be fixed for the duration of the performance. The only exception would be a manually operated lantern such as a follow spot. In the video world it is not uncommon to mount a camera on a motorized pan and tilt head, commonly called a 'hothead', so that the direction in which the camera points can be controlled from a distance. The most obvious use for hotheads is in surveillance, but they also find application wherever it would be inconvenient or dangerous to use an operator. It would of course be perfectly possible to develop a hothead for lighting, but the way the industry has developed is to integrate the pan and tilt mechanism into the luminaire itself. VARI*LITEs use a yoke system where the luminaire itself is free to rotate within the arms of the yoke, as you can see from the photograph (Figure 19.4). Other manufacturers use a moving mirror to reflect the beam from a fixed light into the desired position. The advantage of the yoke design is that the range of angles is very wide. The VL6 spot luminaire, for example, will pan over a full 360 degrees and tilt up to 270 degrees. Although the range of movement might be large, the mass of the entire lamp and lens housing has to move, which means that there is a certain amount if inertia involved, consequently the speed of rotation is specified as 240 degrees per second. Despite the inertia this is quite rapid, if not instantaneous, and the motors must be fairly powerful. If you want faster movement, then you would probably consider a mirror light such as a Martin Roboscan where the moving element is lighter and potentially faster. The price you pay is that the angle of coverage is not as great. Although there are no mirror lights in the VARI*LITE range, they do produce the VLM moving mirror which can be placed in front of a

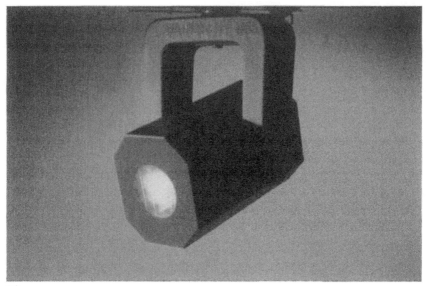

(a)

(b)

Figure 19.4 VARI*LITE VL4™ wash luminaire. (© Lewis Lee.)

conventional lantern to achieve much the same effect at relatively low cost. Although in the past moving lights have been prone to create quite a lot of noise, which limited their use in theatre and television, manufacturers are now keen to ensure that their luminaires are as silent as possible so this should not continue to be a problem. This also applies to electrical noise and Vari-Lite's latest cable system is screened to prevent the high striking voltage of arc lamps from creating a radio frequency pulse that would be very likely to interfere with sound equipment.

Another parameter concerning the directionality of a light is the beam size. A lighting designer may choose to focus the beam tightly into a small area or allow it to spread more widely, in either case with soft or hard edges. The beam size and focus may be controlled by moving the lamp with respect to the reflector, by an iris similar to that in a camera lens or by an internal diffusing mechanism. Once again, in conventional lanterns these would be fixed before the performance whereas with an automated luminaire they can be altered remotely at any time. As an interesting aside, the mirror used in VARI*LITE luminaires to concentrate the beam is a dichroic 'cold mirror' which means that it is reflective at visible wavelengths yet transparent to infrared, therefore much of the heat of the lamp can be dissipated via the rear of the luminaire. There is also such a thing as a 'hot mirror' which will reflect infrared from the path of a beam while allowing visible light to pass through.

To boldly gobo

A lantern has a lot in common with a projector. They both have a lamp, reflector and lens, so it should not be surprising that it is possible to project an image with a lantern. It has been commonplace in performance lighting for many years to put a piece of sheet metal cut out into a pattern into the gate of a lantern to project an image. For reasons I have been unable to establish, this piece of metal is called a gobo. (For that matter, does anyone know why an audio attenuator is called a pad?) Often the gobo would be used to give a soft focus effect to add interest and texture to the lighting. On other occasions it might be thrown into sharp focus to project, for example, a company logo at a sales presentation. Although gobos may still often be low tech metal cut-outs, the production of gobos has been revolutionized by the use of computers and automated machinery. Vari-Lite have a process known as Vari*Image where the gobo is made from specially coated glass onto which the image is laser etched. And unlike metal gobos, this process can achieve half-tones to produce a grey-scale image. Going one step further, an image can be etched onto a dichroic colour glass plate using an ablation process where the metal coating is evaporated away. Vari-Lite claim to be able to produce images from illustrations, floppy disk and even from a fax! Vari-Lite's luminaires incorporate a gobo wheel so that several gobos can be inserted and changed remotely. It is possible to change the gobos continuously, which is known as a rolling gobo effect,

(a)

(b)

Figure 19.5 VARI*LITE VL6™ spot luminaire. (© Lewis Lee.)

and also to stop the gobo wheel halfway between one gobo and another to produce a split beam. In fact this is also possible with the colour wheel, as in the VL6, so that a multi-coloured beam can be created. Although it is not a feature of any of the VARI*LITE range, some luminaires can rotate the gobo to produce a spinning image.

So far I have covered intensity, pan, tilt, colour, beam, focus and gobo change. All of these seven parameters featured in the VL6 wash luminaire can be changed remotely and automated. The only thing that is left to manual adjustment is the lamp alignment, which you would only do when changing a lamp. This means that a single VL6 requires seven control channels from the lighting console. Some luminaires can require over twenty! When you consider the quantity of lights that are commonly used you will see that the lighting operator is in control of far more channels than the sound mixer. VARI*LITES incorporate microprocessors within the luminaires themselves (perhaps we will have microprocessors in our speakers one day) and while some of the VARI*LITE range will run from a conventional lighting console, others need the dedicated VARI*LITE Artisan console. The next time you see moving light and changing colour in live performance or on TV, perhaps you will look out for VARI*LITEs and other competing products and admire the technology involved in performance lighting. It's impressive stuff.

Scanners

Martin Professional, no relation to Martin Audio or Martin guitars apparently, is a Danish company whose range of luminaires includes the Roboscan series and the PAL 1200, all of which create movement through the use of a mirror rather than the yoke assembly of VARI*LITEs. The PAL 1200 offers a 287 degree pan and 85 degree tilt; obviously not as great a range of movement as can be obtained with VARI*LITEs, but the PAL 1200 has other advantages. For instance there is an automated framing system which fulfils a similar role to the barn doors on a conventional lantern, but with very precise control over positioning to a resolution within millimetres, and swivel of up to 22.5 degrees in any direction, and of course it is motorized. The PAL 1200 also offers colour mixing, motorized zoom and focus and provision for four rotating gobos plus a fifth static gobo.

The Roboscan 1220 series, to pick one at random from the Roboscan range, has a high speed shutter that can be used for instant blackout or strobe effects from 2 to 16 Hz. In addition the 1220 series models have rotating gobos plus nine fixed gobos on a gobo wheel. The unit is modular which means that new features may be added to existing units, for example the recent prism rotation option. The modular construction also means that it is possible to carry out repairs such as changing the gobo rotator without having to demount the unit.

Figure 19.6 Martin Professional Roboscan Pro 918 Scanner. (Courtesy Martin Professional).

DMX 512

DMX 512 is the MIDI of the lighting world (and don't forget that MIDI can be used to control lighting as well). The DMX part of the name represents digital multiplex, meaning that many channels can be sent down the same cable. 512 stands for the maximum number of channels on one cable. This may sound very impressive compared with MIDI, but DMX 512's aims and aspirations are quite different and the two are not really comparable. Before DMX 512 (and other protocols of a similar nature), control voltages were sent from the lighting console to the dimmer racks. This, fairly obviously, would require at least one wire per channel plus a zero volts reference. Obviously, a multi-core cable to handle all of these signals would not be practical, so a little digital inspiration is called for.

DMX 512 was designed to be universally applicable, and indeed it has become so and virtually all modern consoles and dimmers speak the same DMX 512 language. A pre-existing electrical interface was used in the design of DMX 512 which called for a single shielded twisted pair cable and a daisy-chain system of wiring with no branches or stubs. This was fine in the days where a fixed installation would have a single cable from the lighting console to the dimmer room, and touring systems were not expected to be particularly complex. This is no longer the case and the simplicity of DMX 512 is now compromised by the need for longer cable runs (than the nominal 1000 feet allowed) and splitters, repeaters and distribution amplifiers.

Although 512 channels probably seemed like overkill when DMX 512 was designed, with modern intelligent fixtures gobbling up 12 to 20 (or more) channels each, the number of lights it is practical to control on one DMX 512 line is limited, according to the mix of intelligent and conventional units, to maybe 30 or 40. The answer, as with MIDI, is for a console to have multiple DMX 512 outputs, and a large lighting console may have as many as six outputs controlling up to 3072 channels. And audio people think 72 channels is a lot!

CHAPTER 20

The art of bluescreen

In both video and film, the storyteller seeks to create an illusion of reality. Strange really, because there is so much on screen – small or large – that is so patently unreal that one has to wonder why sometimes we accept the unreal (editing from one perspective to another, or from one setting in space or time to another), the unfeasible (the craggy fifty-something male lead always getting the girl) and the blatantly absurd (the whole of Last Action Hero). But some things just have to look real, as though what we are seeing really happened right in front of the camera. In the early days of film, special effects were not employed only to astound and amaze the audience, mostly they were done simply to ease the budget – plain old-fashioned story-telling was enough for cinema-goers in those days. One of the earliest special effects was compositing, or the art of combining two images into one, supplementing reality with make-believe to achieve an image that perhaps never could have existed in front of the camera in its entirety. One of the earliest methods of compositing was the glass painting. Suppose for instance that your script demanded a shot featuring the lead characters in the foreground, and an ancient ruined castle in the background. Would you go to the trouble of scouting a suitable location (and there aren't many ruined castles in Hollywood), or would you try and fake it in the studio's back lot? You would fake it of course. The way to do it was to set up a large sheet of glass through which the camera would shoot the scene, including the actors and their immediate surroundings. For the rest of the image, an artist would come and paint in the ruined castle on the glass, and the scene would simply be shot through transparent areas of the glass. It is so simple you would think that it wouldn't work very well – but when is the last time you noticed a glass shot in an old movie on TV? The fact is that you just don't notice them, and they crop up regularly. Glass shots can be used for interiors too. If the film demands a lavish interior setting that would be too expensive to build, just paint it. Painted elements and actual sets can blend together amazingly well, and you have to look for the joins to see them. I imagine that on a cinema size screen these early attempts would be visibly inferior compared to what modern technology can achieve, but in the early days of black and white cinematography they served the industry well.

A glass shot works by combining a still painted image with live action. But what if you need to combine two live action shots together, what then? Well, you might ask why you should do this, since one of the objectives of glass painting was to obviate the need for a special location or elaborate set. In the days before the art of special effects became an end in itself, there was one story device that positively demanded compositing of two live action sequences – creating twins from a single actor. I don't know how many times this device has been used in film and on TV, but it certainly has been well worn in the past, and doubtless will resurface again and again in the future. To combine two live action images, the technique of the matte is used. A matte in essence is a device to block off light from one section of the film so that it remains unexposed. This is sometimes known as a hold-out matte. Subsequently a negative version of the matte can be employed so that the previously exposed film is protected while the new element is shot onto the unexposed section. This would be called a burn-in matte. This can actually all be done in the film camera using a matte box in front of the lens. Shoot twin no. 1 through the hold-out matte, rewind the film and shoot twin no. 2 through the burn-in matte. Maybe it is not quite as easy as having a genuine pair of twins, but technically it is very straightforward, the hardest part being getting the interaction between the characters correct. Often when this device is employed in early film and television, an edge of some kind in the set has been used as the dividing line in the centre of the screen. Since the matte box is closer to the camera than the action, the dividing line between the two sections is automatically blurred slightly, smoothing the transition.

Both the glass shot and the simple matte can be done directly in the camera, but there are limitations to these techniques, and any little thing that goes wrong can ruin the whole shot. In the glass shot, for example, the camera and glass have to be very firmly mounted – any slight movement would soon give the game away. Lighting has to be consistent, especially if outdoors. A cloud passing over the set would almost certainly change the relative illumination on the glass and on the action and make the join between the two very obvious. In glass and matte shots, then one particularly limiting consideration is that in most circumstances the camera has to stay absolutely still: no pan, no tilt, no track and certainly no crane. (Zoom hadn't been invented yet, but that wouldn't have been allowed either.) Camera movements are an essential part of the director's art, and to take away this option severely limits the artistic possibilities of film-making. There has to be a better way, and there is . . .

Rotoscoping

The term rotoscoping dates back almost as far as film itself and originates in a device used to allow cartoon animators to copy live action motion. Call it cheating if you like, but many of the great animated films are packed with scenes that were originally played by live actors or dancers and subsequently reinterpreted (not just redrawn) as animations. These days the term is used to represent any method of drawing something by hand onto film, generally involving rephotographing the hand-drawn sequence, or by drawing or painting via the screen of a computer. But just as you can use the rotoscope, or equivalent technique, to create animation from live action, or even combine animation with live action, so you can use it to create a matte. But rather than the static compositing of the glass shot or conventional matte, now the camera can move. The principles of the static matte still apply, but the methods of application are rather different. One way a so-called travelling matte can be employed is to shoot the foreground element against a background that can easily be identified as such by eye, and then the rotoscope artist can proceed to matte out the background in a process that results in a piece of film with a clear patch where the foreground action takes place, moving and changing shape with it. This is the hold-out matte. The original footage and hold-out matte are assembled into a bi-pack, and exposed via an optical printer onto raw stock. This results in the foreground action being exposed and the rest of the film still untouched by light. From the hold-out matte, simple printing can create a burn-in matte. The background is shot and exposed through the burn-in matte in the optical printer to join foreground and background together into a believable illusion. In fact, the optical printer in its most advanced form can assemble several elements and, amazingly, operates without the benefit of a single bit of digital data in attendance (other than the setting of the on/off switch!).

Rotoscoping is obviously a very versatile technique since virtually any shape of matte you could possibly want can be drawn by hand, but it is slow and labour-intensive. Any process that is slow and labour-intensive is ripe for automation, so visionaries in television and film were considering how this could be done. Skipping over to the medium of television, a simple technique has been used for decades to superimpose titles over live action. Simply create the titles in white letters on a black card and light it evenly. Where the card is black, the video signal is at zero volts and can be mixed with the live action signal which, since it is at a greater voltage, will take precedence. The white letters on the caption will produce a signal close to 0.7 V or peak white which will obliterate the live action. Simple and straightforward, and you can do it with scrolling titles too. As you will appreciate, the trick in this is to have some means of distinguishing between areas that are foreground and areas that are background. In the simple case of titles, all we are effectively doing is saying that one part is black, and that is where the live action should appear on the screen. The other part – the lettering – is white and this is where the live action should be obliterated. This technique has been developed as luminance key or luma key where the signal is switched, or keyed, between foreground and background according to brightness. But luma key has limitations: shooting live action against black is very

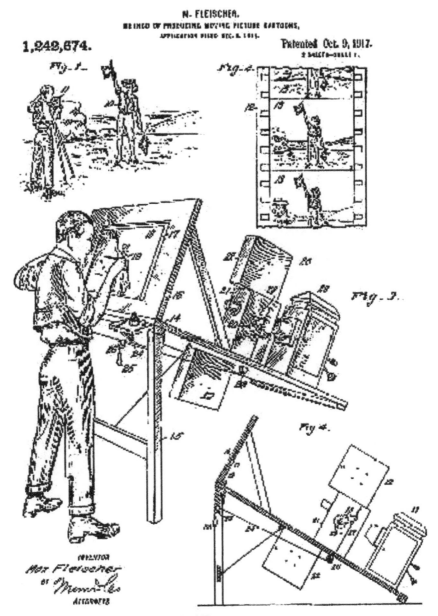

Figure 20.1 Max Fleischer's original Rotoscope.

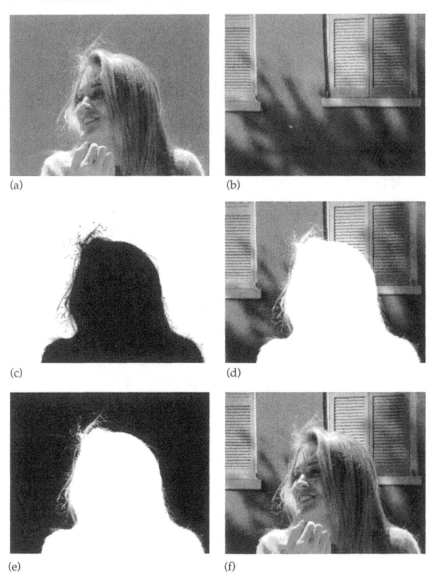

(a)   (b)

(c)   (d)

(e)   (f)

Figure 20.2 The compositing process.

difficult since it is almost impossible to light the foreground without the background greying out. But once colour television was developed another option presented itself. If you shoot against a blue background, the blue colour can be used as the key to operate a switcher. So in the foreground scene, wherever there is blue, the switcher swaps over to an alternative signal. The classic example is the weather report, where the presenter stands in front of a blue screen and the map is keyed in. As you know, this is called chroma key.

Chroma key has been a very versatile technique in television but it has its limitations. One obvious limitation is that there can be no blue in the foreground otherwise the background would appear through 'holes'. Additionally, very careful lighting is required or the blue background will reflect onto elements of the foreground, particularly around the edges, creating the 'matte line' or fringe around the foreground elements that we have all seen so often. In extreme cases, blue will spill into the foreground to an extent that holes appear again where the keyer sees excess blue. Fringing can be reduced by careful backlighting with a yellowish colour that cancels out the blue. Chroma key has been developed over the years and there is no longer the hard-edged separation between foreground and background, and inability to resolve fine detail such as hair, that there used to be. But the fact remains that good old-fashioned chroma key is just that – it switches to one signal for the foreground, to another for the background. On/off, either/or. But there's a better way . . .

Ultimatte

Chroma key was a useful intermediate stage of development in compositing, but it had obvious limitations. There were numerous advances over the years but the ultimate was, well, Ultimatte. The difference between simple chroma key and Ultimatte is that Ultimatte is not simply an on/off device. Rather than seeing the dividing line between the foreground object and the blue background as an edge, Ultimatte looks upon it as a gradual transition. The Ultimatte algorithm is therefore capable of resolving fine detail such as hair and smoke. Smoke is a particular problem with simple chroma key. Where it is dense enough to obscure the background it looks fine, but where the smoke is thinner the background shows through, resulting in a very visible dark edge. Ultimatte also considers the brightness of the blue background. Where the blue background is darker, then the inserted background image will be darker. This means that where a foreground object or actor casts a shadow on the blue screen, that shadow will be imprinted on the background image when the composite is made. Fairly obviously, this leads to the requirement that the blue background must be very evenly lit, but it is possible to capture a still image of the background, and automatically correct for unevenness in the lighting for subsequent shooting, although it has to be said that creating and lighting an even background is still a very important part of the bluescreen art. Although modern high-end chroma key units can achieve results that can approach Ultimatte, Ultimatte has another feature – blue spill removal. Spill from the blue background is quite likely to reflect onto the foreground action and at worst can cause holes, as previously mentioned, while at best the blue is visible. But with Ultimatte, blue spill can be compensated for, resulting in a very believable composite with no obtrusive artefacts.

Swapping back to film, since film is not an electronic process there is no equivalent of chroma key, although bluescreen techniques can still be applied in a different way. In traditional film technique, the foreground action is shot against a blue screen. The result can be used to generate a matte by optical means. In this case the matte does not have to be just clear or black, like a rotoscoped matte would be. The matte can have a varying density and capture fine detail around the edges of the foreground quite well. A variable density matte is sometimes known as an alpha channel. To do this optically requires a very even blue screen precisely exposed and is difficult to the point of near impossibility. Fortunately, digital technology is at hand, and with digits things do not have to be quite so accurate at the shooting stage as a wider range of compensation can be employed later on. For instance, the matte does not necessarily have to cover the entire area of the image any more as long as the action is clearly differentiated from the background. A 'garbage' matte can be digitally painted in to cover the rest. (In fact this technique is also available optically, but the matte has to be very carefully lit to prevent its edges from showing.)

Compositing for film is now done in the digital domain and optical techniques have all but disappeared. Television and film technology have been drawn together and similar techniques are available in both media. Of course, at the lower resolution of television, and considering that it is an electronic signal, real-time processing is standard and systems such as Ultimatte exist as units with knobs to adjust the process parameters. You see the results of your work immediately and any lighting deficiencies can be taken care of. In film, you don't see the results until the negative has been scanned into a Silicon Graphics, NT or Macintosh workstation running compositing software (or plug-in), but the film production process tends to operate on a longer timescale so if there are any problems they can be dealt with using the digital equivalent of rotoscoping and making corrections directly on the screen.

Motion control

I have left one important element of compositing unexplained. Travelling mattes are all very well, whether rotoscoped or generated automatically, but they only work if the relative perspectives between foreground and background match, and even in a locked-off shot this is difficult to achieve. When the camera moves then matching the change in perspective between foreground and background is essential otherwise the composite just will not be believable. There are exceptions, such as where the foreground action takes place against a background in which there is no obvious perspective reference. But if the foreground and background are closer, or if in the extreme they appear to be entwined into each other, then when the camera moves in the foreground shot, the perspective of the background has to change to match it, and they will have to continue to match perfectly from frame to frame. This has in the past been accomplished by mechanical means, but now of course the computer is in charge. Motion control photography is in essence a pre-programmed series of camera moves including pan, tilt, track, crane and zoom. Of course the modern language of cinematography often requires a combination of these to happen all at the same time, so the camera is installed on a motion control rig and a pre-programmed series of moves is carried out to follow the foreground action. Separately, that series of moves is replicated exactly for the background. As you are probably aware, backgrounds are commonly created as miniature models, so the motion control computer must be able to scale movements down to size, and obviously a great deal of optical expertise will be necessary too. One further advantage of motion control is that when using models, it is possible to do a separate pass purely to generate the matte. In conventional bluescreen work, one of the major problems is lighting the foreground artistically while at the same time lighting the blue screen sufficiently evenly. With motion control, lighting can be adapted purely to the requirements of the matte, and therefore a better finished product is achieved more easily. Next time you go to the cinema, or even watch an old movie on television, watch out for compositing. Not only does it allow the filmmaker to get more out of the budget, but it makes the impossible possible on the screen.

Blue screen or green screen

The reason why compositing techniques have traditionally used a blue screen for the background is simply because blue paint of sufficient quality and consistency was easier to come by than green. That problem has now disappeared and blue screen or green screen backgrounds are used according to the requirements of the job. For instance, it wouldn't be much good trying to shoot Superman flying against a blue background. His costume is blue and would be matted out of the foreground leaving only his head, hands, feet, red cape and yellow 'S' logo. Likewise, if it proved necessary to composite the Jolly Green Giant against the Manhattan skyline then a blue screen background would be essential. Where that leaves the Teletubbies is open to debate! The reason why red is seldom used for compositing or keying is because skin tones often contain a lot of red, and even if they don't, then pink or red lipstick or makeup would be out of the question. Red screen is used mainly for advertising where products or packaging are shown that incorporate blue or green hues, and obviously it is not possible to change corporate or brand colours to suit the technical requirements of the compositing process.

APPENDIX 1

The science of colour

Suppose you were in contact with a being from another world via a hyperspace postal link. How would you show your two-headed friend what our world looks like? One way might be to get an extra set of prints from your holiday photographs and beam them over. But would he, she or it see the photos the same way you do? Probably not. Our eyes are tuned to the narrow range of electromagnetic frequencies that are useful for vision and can penetrate the Earth's atmosphere. What's more, within the range of visible wavelengths available covering nearly an octave of frequencies, the retina only has sensors for three colours – red, green and blue. From this limited information we form an opinion of what our world is like. Our colour photographs and colour televisions are designed to exploit the limitations of the eye and produce a picture whose colours re-create passably well those of the real world. In actuality, however, the colours on a photographic print may be nothing like the colours that originally existed in front of the camera, and our alien friend might get a big surprise if he thinks that our planet is going to look exactly like the photos when he calls by in his flying saucer. His eyes will in all probability be sensitized to a completely different quality of light and the colour prints won't fool his vision the way they do ours.

Light

An easy answer to the question 'What is light?' would be, 'The part of the spectrum of electromagnetic radiation that we can see.' Electromagnetic radiation is the cover-all name for radio waves, microwaves, infrared, visible light, ultraviolet, x-rays and gamma radiation. They are all the same thing, only the wavelengths – and therefore the frequencies – are different. Figure A1.1 shows the small part visible light has to play in the scheme of things, just an octave of frequencies – almost – in a range which covers 10^5 Hz to 10^{25} Hz and beyond. Much of the longer wavelength range of electromagnetic energy emitted by the sun gets through the atmosphere, but it would be no good to use radio waves for vision since their wavelengths are far too large to resolve small objects, and we would

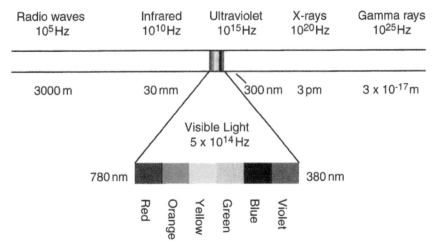

Figure A1.1 The electromagnetic spectrum.

need eyes the size of radio telescopes. Ultraviolet radiation, as you know, is stopped to a large extent by the ozone layer, what little is left of it, and we are largely insulated from shorter wavelengths too.

White light consists of all the frequencies within the visible range, and you can split it up with a glass prism or diffraction grating into a spectrum of colours. We see bands of colours and name them red, orange, yellow, green, blue and violet, but this is just a trick of the eye. The spectrum is continuous and each colour blends gradually into the next. The range of wavelengths that the eye can accommodate is approximately 400 nanometres (nm) to 700 nm.

The eye

The retina of the eye is covered with light-sensitive cells called rods and cones because of their shape. The rods are sensitive only to the intensity of the light falling upon them and the cones are responsible for colour vision. Although the cones are responsive to colour, they are ten thousand times less sensitive than the rods, so colour vision diminishes when there is only a small amount of light available. Cones come in three varieties: blue-sensitive, green-sensitive and red-sensitive; and their relative responses, peaking at 430 nm, 560 nm and 610 nm respectively, are shown in Figure A1.2.

Light consisting of a single wavelength is called 'monochromatic' and evolution could have designed an eye so that there were groups of sensitive cells, each group being responsive to a very small band of

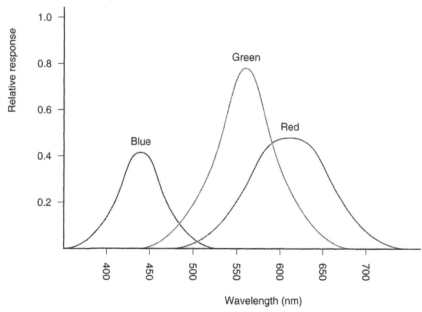

Figure A1.2 Relative sensitivity of the red, green and blue sensitive cells in the retina.

wavelengths or frequencies. This is pretty much the way the ear works, and if the eye were the same then we could say that we had true colour vision. Unfortunately, if this were so, adequate colour television would still be an impractically complex dream so perhaps we are better off as we are. We do, however, see colours other than red, green and blue, and this is because, for instance, when our red-sensitive and green-sensitive cones are stimulated simultaneously, the brain interprets this as yellow. Our television sets therefore do not need a yellow phosphor – a combination of red and green phosphor dots glowing simultaneously and very close together will be seen as yellow. It is perfectly possible to reproduce closely many of the colours of the spectrum by blending together different proportions of red, green and blue, or other spectral colours. But there are also colours in nature which are not in the spectrum of colours produced by a hot glowing object such as the sun. The various shades of purple are examples of non-spectral colours and are produced by combinations of red and blue light, which are at opposite ends of the spectrum. Looking at Figure A1.2 it is evident that there is no spectral colour, monochromatic and of a single wavelength, that can stimulate the red and blue sensitive cells simultaneously without also stimulating the green. We need the combination of two distinct spectral colours to achieve this, and it happens often in nature to produce the colour we call purple.

Subtractive colour mixing

We are intuitively most familiar with this form of colour mixing since this is how we did it as children, messing about with pots of paint. A leaf appears green under white sunlight because it absorbs, or subtracts, all the other colours from the light and reflects only the green. Paint pigments do the same and artists talk about the primary colours, blue, red and yellow, from which other colours can be made. Photographers or printers would know these subtractive primary colours by their correct names of cyan, magenta and yellow. A cyan pigment absorbs red light and reflects the rest, magenta absorbs green and reflects the rest, yellow absorbs blue and reflects the rest. So if you mix together yellow and magenta pigments then effectively you are subtracting blue and green from white light leaving only red to be reflected from the white paper. Similarly, mixing yellow and cyan produces green, and mixing magenta and cyan makes blue. Mix all three together, and if they are accurate and pure subtractive primary colours, then the result will be black since all the light falling onto the paper is absorbed.

Subtractive colour mixing is interesting, but only useful if you are mixing pigments and painting them onto paper or canvas, or making colour photographic prints. More interesting to us is the alternative technique of additive mixing which is used in colour television.

Additive colour mixing

With additive colour mixing, the three primary colours, from which all others are made, are the familiar red, green and blue. If you had three flashlights and covered the lenses with red, green and blue filters then you would find that:

    blue + green = cyan
    red + blue = magenta
    red + green = yellow
    red + green + blue = white

This assumes of course that you are shining them onto a white surface in a darkened room. In additive mixing, cyan, magenta and yellow are known as complementary colours. If a complementary colour is added to the colour that it does not itself contain, in the correct proportion, then the result will be white. Mixing coloured lights shining onto white surfaces is one way of adding colours. Having many very small coloured lights clustered together which are viewed directly is another. Look very closely at your television set (not for too long please) and you will see that this is exactly what the screen of a television set is.

Colour triangle

The colour triangle shows the relationships between the three primaries and allows any colour to be described in terms of numerical co-ordinates. This is the system devised by the Commission Internationale de l'Eclairage and is known as the CIE Colour Triangle. This diagram (Figure A1.3) contains all the colours it is possible to produce from three primaries. At the corners are the three primaries, red, green and blue (actually versions of these colours that cannot exist, but theoretical 'super-saturated' primary colours that are necessary for the creation of this triangular concept). In the centre, where all the colours mix together, are the various possible shades of white. And if you thought that there was only one white, think again. There are several 'standard' whites in use as I shall explain more about shortly.

If the natural world were a little more straightforward than it is then real colours would exist all the way to the edge of the outer triangle. Along the angled side of this right-angled triangle we would find orange, yellow and yellowish green. These colours would be produced by combining two of the fictitious super-saturated primaries necessary for the concept that are redder than red, greener than green and bluer than

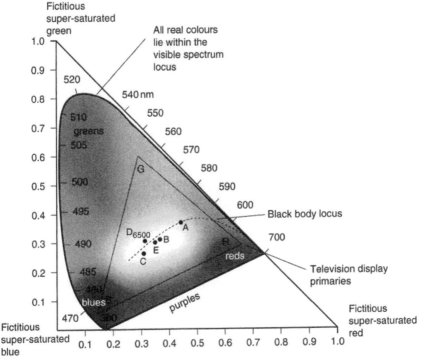

Figure A1.3 CIE chromaticity triangle.

blue, if you see what I mean. Midway along the base there would be magenta and halfway up the side would be cyan. The horseshoe shape within the outer triangle is a little more representative of the real world. These are all colours that actually exist and can be represented by combining primaries. Along the curved section are all the spectral colours that can be imitated by combining various proportions of the fictitious super-saturated primaries. The straight section along the bottom is where red and blue are combined to produce the non-spectral colours, the various shades of purple. The true spectral colours, which are obtained by splitting up white light, are, virtually by definition, saturated. This means that they contain no white. The further away a colour is from the edge of the horseshoe, the more white it contains and the less saturated it is. Another word, hue, describes the position of the colour around the edge of the horseshoe. Hue and saturation are used a lot in colour television terminology. For example in the NTSC system, to a moderately close approximation, the phase angle of the chrominance signal describes a colour's hue, the amplitude of the chrominance signal describes its saturation.

You will also notice a smaller triangle within the horseshoe and outer triangle. This represents the range of colours obtainable on a TV set. A little less than half the range of colours available in the real world, I would estimate. The positions of the television display primaries are governed by the nature of the phosphors used. In European colour TV systems, other phosphors are used which give a rather brighter image than NTSC, but the gamut, or range, of colours is not as wide. Even if the phosphors of a TV set were perfect, then this system of combination of primaries still could not reproduce all the colours, the full gamut, of real life. Take a look at some flowers, or a butterfly's wing, and you will see colours that will never appear on TV or on the printed page.

Colour temperature

If you have ever been involved to any depth with photography or video then you will have come across this term. It describes the precise colour of 'white' light used to illuminate the scene to be shot. As you know, the light from a tungsten filament bulb is much 'warmer' than the bluer tone of daylight. In photography, you have to choose a film specially balanced for tungsten light, or use a daylight colour film with a blue filter on the lens. In video, you have to set the white balance on the camera before shooting. Many cameras will estimate this automatically. In television, there has to be a standard white or viewers would see different colour balances on different TV sets (I know they do already, but the problem would be even worse without a reference point!). Colour temperature is a theoretical concept with practical ramifications. Let me start with the theoretical side of things.

Suppose you had an object which absorbed all electromagnetic energy falling upon it. You would call this, as physicists do, a black body. Suppose this black body is made of a fairly robust material that doesn't melt, boil or generally fall apart when heated, then as its temperature increases it will start to glow, first dull red, then bright red, then reddish white and finally bluish white. This is the 'black body locus' shown within the CIE triangle. The section where the black body is reddish and bluish white is of most interest since it approximates to the different shades of white that we are likely to come across. To define the exact shade of white all you have to do is measure the temperature of the black body in Kelvin (which is the same as measuring it in degrees Celsius and adding 273). For example, if you heat a tungsten filament to 2800 K then it will emit light with a colour temperature of 2854 K. The reason why these figures are not the same is that a tungsten filament is a close but not an exact replica of a theoretical black body.

To go with the concept of colour temperature, there are five 'standard illuminants' in common use which are known as Illuminants A, B, C, D6500 and E. Their positions are shown on the CIE triangle. Illuminant A is the previously mentioned tungsten lamp heated to a temperature of 2800 K and emitting light with a colour temperature of 2854 K. This would be considered a fairly 'warm' white. Notice that the point appears exactly on the black body locus whereas the other standard illuminants do not exactly match the colour that a theoretical black body would emit. Illuminant B is a very close match for noon sunlight and has a colour temperature (strictly speaking a correlated colour temperature since it doesn't exactly match a black body) of 4800 K. Illuminant C is similar to the light from an overcast sky and has a correlated colour temperature of 6770 K. Illuminant D6500 – the one with the long name – has a correlated colour temperature of 6500 K, as you guessed, and is the white that is aimed at in colour television displays. (But not always, because lower colour temperatures are used for TV sets that are used as part of a production and actually appear on camera. Higher colour temperatures are sometimes used to achieve a more 'punchy' picture.) Illuminant E is a theoretical light source which would exist if all the wavelengths of the visible spectrum were present with equal energy. The visual equivalent of white noise if you like.

Looking into the future, it seems that the colour triangle will be with us for years to come. Somewhere deep in a secret research laboratory perhaps someone is struggling to devise a display where the hue of a colour will be modulated directly rather than being crudely simulated as it is now. Or am I just dreaming?

Light terminology

Luminous intensity is the power of a light source. The unit is the candela (cd) and is 1/60th of the light emitted from 1 cm^3 of a full radiator ('black body') raised to the melting point of platinum, 2042 K.

Luminous flux is a measure of the rate of flow of light energy. The unit is the lumen (lm) and is the luminous flux emitted in one second per steradian (solid angle) by a point source of 1 candela. A point source is an infinitesimally small sphere which radiates light equally in all directions. Another way of describing the lumen is the quantity of luminous flux per second which passes through a one square metre area of a transparent sphere of radius one metre with a point source of intensity one candela located at the centre.

Illuminance is the concentration of luminous flux striking a surface and is measured in lux. 1 lux = 1 lumen per square metre.

What colour is grass at night?

This is a question to tax philosophers. Some might say that grass is intrinsically green and we only need light to observe that greenness. Others would say that the colour green depends on the existence of light and until it is illuminated the grass remains as black as night. Doubtless the answer to this question will remain the subject of heated discussion (somewhere) but a more practical question relates to the colour of grass under sodium vapour lighting. Sodium vapour street lights emit a very narrow range of colours – in fact there are only two yellow components, very close together in the spectrum. This light strikes the grass and is reflected back from the green pigment. The green pigment can only reflect green and has no power to change the colour of the light, therefore what tiny amount hasn't been absorbed will still be yellow, but a very dim yellow indeed. My opinion is that grass at night under sodium lighting is a slightly yellowish black. What do you think?

Why do compact discs seem to have 'rainbow' colours on the playing surface?

It is well known that a glass prism can split white light up into its component colours. A pattern of very fine lines inscribed on a transparent or reflective surface can do exactly the same and is known as a diffraction grating. The surface of a compact disc doesn't have lines, but the spiral pattern of tiny pits containing the digital audio data amounts to pretty much the same thing. We don't have to wait until it is raining to see a rainbow – isn't technology wonderful?

Timecode: the link between sight and sound

Timecode was developed to solve a particular problem – that of editing video tape. Before video, all programmes were live or on film. Video recording was first seen as a means of conveniently time-shifting live programmes for broadcast when people, in their various time zones, were likely to be home from work and still awake. A step forward in thinking from time shifting is the possibility of showing a programme more than once. But this is hardly stretching the concept of video recording beyond being a mere convenience.

Sound editing had been used as a production tool since the days when radio programmes were recorded on 78 rpm discs. Dub edits, from disc to disc, were made to assemble a complete programme. But once video recorders were up and running, it must soon have been obvious that if it were possible to edit a video tape, TV programmes could be assembled in a similar way. The problem is that a video recording is discontinuous – a series of individual frames – so it is not possible to cut and splice the tape or perform dub editing in the same way as you can with audio tape. Whereas sound edits have to be done with a degree of precision, a video edit has to be done at exactly the correct point between the frames, maintaining exact sync, or the picture will break up. But even if it was going to be difficult, the early workers in video were determined to be able to edit recordings, and the first method was literally to cut the 2-inch tape and splice it back together again. Imagine doing that with your VHS tapes. The splice points could be found by making the video waveform – not the image – visible by painting a suspension of fine magnetic particles onto the tape. Later on, electronically controlled dub editing, from machine to machine, was made possible by recording a series of control pulses onto a separate track on the tape. Two machines could be synchronized using these pulses, but there were problems due to tape drop-outs and the fact that the pulses did not identify the exact frame. Edit points were approximate at best.

In 1967, the Society of Motion Picture and Television Engineers (SMPTE) devised a system which would label each frame with a unique code which came to be known as SMPTE timecode. Timecode is a sequence of pulses, digitally encoded with time-of-day information, which can be recorded onto an audio track of the tape. By this means an electronic video editor can make frame-accurate joins. This technique is still in use today, and its use has been extended to all the timecode procedures we currently employ. So you see, timecode was invented purely as an aid to video editing. They weren't thinking about audio at all.

The nature of timecode

Timecode starts life as electrical pulses created by a timecode generator. The generator converts these pulses into a waveform which looks, to a tape recorder (or the audio track of a video recorder), just like a typical audio signal. To edit video recordings successfully timecode must contain two distinctly different types of information. Firstly, timecode should contain a clock pulse so that, by means of comparing the clock pulses from two tapes, two machines can be made to run at exactly the same speed and never drift apart. Secondly, the timecode should identify each frame of a reel uniquely so that the synchronizing and editing equipment always knows where it is in the reel. Film is a useful analogy where the sprocket holes, four per frame, can be used to provide a clock pulse, and numbers along the edge of the film measure the film's length in feet. There are a number of ways in which the timecode signal could be created:

- Return to zero (RTZ) code where positive and negative pulses below and above a reference voltage indicate binary 0 and 1. It is self-clocking but vulnerable to accidental polarity inversion when the 0's and 1's will be swapped.
- Non-RTZ code with two states indicating 0 and 1. Non-RTZ code does not contain a clock and is also polarity conscious.
- Frequency shift keying (FSK) where two different frequencies are used to represent 0 and 1. Requires a relatively high bandwidth to record and cannot be read at anything other than normal play speed.
- Biphase mark. Contains a clock pulse, is immune to polarity inversion and can work over a wide range of play speeds. Perfect!

In biphase mark code, each timecode frame (which in a video recording corresponds to a single frame of video) contains a 'word' consisting of 80 bits. For each of these bits, the voltage of the signal changes from high to low, or from low to high. If a bit is to represent a binary 1, then there is a second transition halfway through the bit period. If the bit is to be a binary 0, then the voltage remains constant until the start of the next bit period. Figure A2.1 shows a sequence of timecode bits, and their binary values. Notice that it is not the absolute voltage level that determines the value of each bit – it is how often the level changes. This means that if the

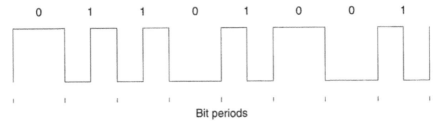

Figure A2.1 Biphase mark code. The level changes every bit period. If there is no transition during the period then a binary 0 is encoded. If there is a transition halfway through the bit period then it is a binary 1.

timecode signal is accidentally phase reversed, probably through the incorrect wiring of a connector, it still has the same meaning. It also means that the information can be read both forwards and backwards. If running timecode backwards is not something that forms part of your daily studio routine, rest assured that all will be explained in due course.

Table A2.1 shows the meaning of each of the 80 bits in the timecode word. As you can see, the content is simply time information, plus user

Table A2.1 SMPTE timecode data structure.

Bit	Use
0–3	Frame units
4–7	User bits group 1
8–9	Frame tens
10	Drop frame flag
11	Colour frame flag
12–15	User bits group 2
16–19	Second units
20–23	User bits group 3
24–26	Second tens
27	Biphase mark correction
28–31	User bits group 4
32–35	Minute units
36–39	User bits group 5
40–42	Minute tens
43	Binary group flag 0
44–47	User bits group 6
48–51	Hour units
52–55	User bits group 7
56–57	Hour tens
58–59	Binary group flags 1 and 2
60–63	User bits group 8
64–79	Sync word

bits (which may be used for station or reel identification, etc.), plus a couple of flag bits which I shall come on to in due course. The sync word is a sequence of bits which is used to maintain synchronization and to determine whether the tape is travelling backwards or forwards. So you can see that there are two distinct functions carried out here. One, as I said, is to provide a consistent clock pulse, the other is to convey time of day information. Timecode is really rather clever stuff, and not just a horrible screech that offends our ears.

Types of timecode

Although the USA has been fortunate in getting other parts of the world (particularly England!) to use their language, they were not so fortunate as to get their 60 Hz mains frequency adopted as well. For kettles and hair dryers this doesn't make any difference (although the voltage difference does), but for early TV systems the frequency at which the mains voltage alternates provided a very convenient frequency reference. Thus in the USA, TV has a frame rate of 30 Hz (half the mains frequency). In Europe and many other regions, the frame rate is 25 Hz. Since timecode provides an identification for each frame of the picture, as well as a sync reference, it follows that US code must count up to 30 frames, European code must count up to 25. Therefore there must be at least two incompatible types of timecode. In practice there are six(!) and since video plays to a world market a little understanding is necessary.

People working for the US market juggle awkwardly between 30 fps (frames per second) code and code which runs at 29.97 fps. 30 Hz was the frame rate before colour TV was invented and unfortunately was incompatible with the frequency of the sound carrier in the NTSC colour system, so the frame rate had to be reduced slightly to 29.97. All types of timecode identify frames in hours:minutes:seconds:frames. Frames can be further divided into subframes by interpolation within the timecode reader, but subframe information is not contained within timecode itself. A typical timecode display might show 01:01:01:01.25 which translates fairly obviously as one hour, one minute, one second and one and one quarter frames.

As I said, timecode itself identifies frames, not fractions of a frame. This is fine for 30 fps code as the frames can be counted up to 30 and then the seconds counter is incremented. 30 whole frames make up one second. This does not quite work for 29.97 fps code since the tape is running slightly slower. It would not make sense to give frames fractional numbers, neither would it be possible to count . . ., 26, 27, 28, 29, 29.97, . . . so a little bit of trickery is called for. Let me skip sideways to the subject of leap years which is a very close analogy. As you know, the Earth takes 365 days plus one quarter of a day to travel around the sun, yet a year only has 365 days. Over a period of decades and centuries the seasons would gradually slip so that eventually summer would come in December and winter would come in July, in the Northern hemisphere. So we add an extra day every four years to make up for four quarter days and we call it a leap year. But . . . that quarter day is not actually an exact quarter, it is slightly less than a quarter, so adding an extra day every four years doesn't quite work. To compensate for this, every century year, which as it is divisible by 4 would normally be a leap year, is not a leap year. Hence the year 1900 was not a leap year although 1896 and 1904 were. Even now, the compensation is not perfect, so every century year that is divisible by 400 actually is a leap year. So in the year 2000 there was indeed a February 29.

It is the same with 29.97 fps timecode. Frames are still counted up to 30 in each second, but compared with the clock on the wall the timecode clock will run slow. To compensate for this, two frame numbers are skipped every minute, except every tenth minute when they are not skipped. This is called 'drop frame' timecode. When we talk about 'drop frame' code, however, we must remember that no frames are lost – only the numbers. It should really have been called 'skip frame number' code but that is a mistake of history. Despite these periodic corrections there is still a residual error of 86.4 µs over a twenty-four hour period. This is surprisingly important and timecode generators used by broadcasters can correct this each day at a time that will cause least disruption to production and programming.

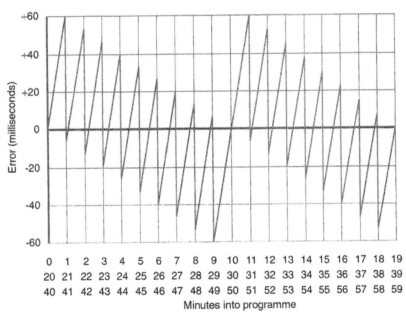

Figure A2.2 Cumulative error in 29.97 fps drop frame timecode.

29.97 fps non-drop frame code and 29.97 fps drop frame code are both widely used in production, although by the time the master tape is completed it should be given 29.97 fps drop frame code. The advantage of working with 29.97 fps non-drop frame code is that the numbers are simpler to work with, although you need to be careful about the duration of the programme. With 29.97 fps drop frame code, running times are correct but you need to maintain an awareness of the missing frame numbers. To make matters worse, people working in film and audio-only production commonly use 30 fps non-drop frame timecode, and in film intended for transfer to video 30 fps drop frame timecode can be used. 24 fps timecode for film is now rarely seen. I reckon that a significant proportion of employment in the audio for video industry is in correcting mistakes made by using the wrong type of timecode, or by assuming that the timecode on a tape is one type when it is actually another. It's a nightmare scenario already and I haven't even mentioned pullup and pulldown.

North Americans who are struggling with the problem could of course consider transferring operations to Europe where timecode is 25 fps and drop frame timecode doesn't exist. The only problem is that much of the programming originates in the USA and is sent over with 29.97 fps drop frame timecode ... or was it non-drop frame, or was it 30 fps. At least material going from Europe to the USA will always (hopefully) have the same 25 fps code.

Timecode generation

The usual source of timecode is a timecode generator. Quite often equipment principally designed for other purposes contains a timecode generator but probably without the full range of facilities you might need. For broadcast applications, and in other types of production too, the generator is set to produce timecode with the real time of day, so when you press the red button, timecode is recorded starting from the actual time you started rolling. Elsewhere, a timecode generator may be set to start running from 01:00:00:00 or 10:00:00:00. The big no-no is to start from 00:00:00:00 because you never can tell whether at some later stage some new material may be inserted before the existing material, and would therefore require a timecode earlier than 'midnight'. Timecode that crosses midnight can confuse a synchronizer or timecode reader and is therefore to be avoided. Starting from one hour or ten hours means that you can easily calculate where you are in the programme with just a little mental arithmetic.

For purely audio purposes, the timecode can be free running without any external reference. Normal practice, as you know, is to stripe one track of an analogue tape from beginning to end with code before any programme recording takes place. Digital tape may be striped, or it is often possible to derive SMPTE timecode from the format's internal timing reference. Since timecode is recorded on analogue tape in the same way as any other audio signal, it follows that you have to set the correct level. Getting a good signal-to-noise ratio, in this case, is usually not as important as maintaining a good quality of code. Timecode, as a pulse waveform, is full of sharp corners which need to be well recorded to ensure correct decoding. Sharp corners on a waveform always indicate strong high frequency content. Analogue tape recorders, although they may extend beyond 20 kHz in their high frequency range, do not maintain their performance at all recording levels. The higher the level (and the slower the tape speed), the harder it is to record high frequencies. Timecode, therefore, should be recorded at a few decibels below 0 VU. The only way to find the correct level is by experiment but somewhere between –5 VU and –10 VU should work. Recording timecode onto analogue multi-track tape throws up another problem – crosstalk. The sound that timecode makes could not be more objectionable to the ear. Even the tiniest little bit leaking into the adjacent track, or even the track beyond, can make a recording unusable. A recording level must be found which enables correct decoding, yet keeps crosstalk to a minimum. On analogue multi-track tape, timecode is almost always recorded on the highest numbered track, which means that there is only one adjacent track for the crosstalk to leak into. This raises the interesting point that in the days before timecode, engineers were always advised not to put anything important on the vulnerable edge tracks of the tape, and now what do we do but put our all-important sync reference on an edge track! There is a wider tolerance of levels when striping timecode onto digital tape, and fewer crosstalk problems, although a new priority is to synchronize the timecode to the digital data stream in the same way that timecode must be synchronized to a video signal.

Recording timecode onto the audio tracks of analogue video recorders is more difficult than on audio machines, which is why an industry has grown up producing timecode 'fix-it' boxes of all types. Audio tracks on video recorders have until recently been second best to the picture, which is one reason why we have had to sync up dedicated audio machines in the first place. But if these slow-moving skimpy tracks are not very good for audio, can they be good for timecode? The answer is no, but fortunately there are techniques to provide a remedy to most problems. Reading timecode should in an ideal world be a straightforward matter, but often it is not. Unfortunately, even with modern tape formulations, drop-outs still occur. And a drop-out on the timecode track means a momentary loss of synchronization. Timecode readers – which may be separate units or incorporated into the synchronizer – come in all quality levels from 'give me good code or I won't play with you any more' up to sophisticated units which can cope with very badly mangled code. Basic timecode readers need code with good sharp edges and free from drop-outs. If the code disappears, even momentarily, sync will be lost. Better readers will cope with code with rounded edges and jitter, such as you are bound to get from the audio tracks of VHS recorders. Also there may be a 'flywheel' feature where the reader will fill in for any missing or damaged code until good code is found again.

Nevertheless, however competent the reader is, there is no substitute for having good code in the first place. If Rule 1 is to record at the correct level determined by experiment, Rule 2 is never to copy code without regenerating it. It is often necessary to copy code, but if this is done by hooking a cable between audio out and audio in from one machine to another, the nice sharp edges of the code will be lost. The only correct way to copy code is to regenerate it, which means passing code from the playback machine through a device (probably incorporated in the timecode reader or generator) which will decode the timecode numbers, make sense of them and compensate for any discrepancies, and then reissue perfect code to the recorder. A moderately satisfactory alternative is to reshape the code. This means that the sharp edges are restored, but drop-outs and errors are not compensated for.

Sound and picture, working together

I said I was going to start at the simple end and you can't get more simple than the system used for synchronizing a film and its sound track. If film sound suddenly became de-invented and someone asked you to develop a system for synchronizing the sound and the picture it probably wouldn't take you long to come up with the idea of recording the sound next to the images on the same piece of film. Take this a step further and you can have two lengths of film, one with picture and one with sound, which have exactly the same dimensions. And what about those sprocket holes (Figure A2.3)? It wouldn't be hard to rig up a mechanical device which transported both films at the same speed, would it? Sprocket hole synchronization is virtually foolproof, and when it comes to editing the film all you need is to have a reference mark on both picture and sound films, generated by the clapperboard, and to make sure that you cut the same lengths of picture film and sound film. Of course that doesn't explain the art of film editing but the method is straightforward.

Of course, sprocket hole sync depends on being able to record sound onto a piece of film, either magnetically or optically. Recording sound on film has its advantages but it has its disadvantages too. It is not really practical to shoot sound onto mag film since the machinery is far too bulky and quarter-inch analogue tape has been the standard medium of acquisition for many years, and is still current. But now of course the tape has different physical dimensions to the film and you will undoubtedly notice the lack of those handy sprocket holes! To overcome this problem the Pilotone and Neopilot systems were developed which, in essence, take a signal generated by the camera from the film's sprocket holes to

Figure A2.3 In traditional film technique, sound is recorded on magnetic film with the same dimensions as the image allowing simple mechanical synchronization.

record a pulse signal on the tape. On transfer to mag film in the traditional way of doing things, this pulse could be used to regulate the speed of the film sound recorder. This would result in a synchronized sound film with a firm sprocket hole link back to the original camera film.

Although sprocket hole sync and Neopilot can keep two machines running together at the same speed since they both provide what we would now recognize as a clock pulse, neither method provides any positional information so the picture and sound could be running at precisely the same speed but ten frames adrift. In film, the position reference is generated by the clapperboard, as mentioned earlier, which provides distinct audio and visual cues. Further down the film, frames can be identified by edge numbers (or key numbers) which are copied during duplicating operations and are useful to editors. Simple though film synchronization is in theory, it has developed into something more capable and versatile, and inevitably more complex.

Finger sync

Back in video-land, let us assume that you are a music composer just about to get your first break into television and you already have a MIDI set-up in your home studio. What additional equipment do you need – absolutely need – to start work? I should maybe have put the answer at the bottom of the page and asked you to write a list because it is a lot simpler than many people think. All you need to write music to picture is a stopwatch, as simple as that. That's the way they did it in the old days and many a wonderful film score has been composed with no more complex equipment. The modern version of the stopwatch is burnt-in timecode and a timecode calculator. Burnt-in timecode, which is explained later in a panel, is where the SMPTE numbers identifying hours:minutes:seconds:frames are branded onto each frame on a copy of the edited programme. Using the still-frame mode of a domestic VHS recorder the composer can read the timecode values of the start and end of a scene, enter them into the timecode calculator and read out the exact duration of the scene, then away to the MIDI sequencer.

So far so good, but what happens when the composer wants to try out the music against the picture to see how well it works? How can the sequencer be synchronized to the video? The answer is to press the play button of the video, or its remote, with one finger, and press the start button of the sequencer with another – simple 'finger' sync. Although the video will take a little longer to burst into life, you can get the hang of the relative response of the two systems and it can be a very effective way of working. Before the days of MIDI sequencers when recording would have been to multi-track tape I know of one in-house music production centre of a major broadcasting organization that never used any other method of synching audio to picture during the multi-track recording process. Finger sync can therefore get you started as a TV composer, but what's the next step?

Code-only master

The next step up from finger sync is to use your home VHS as a code-only master. Amazingly enough, as it comes from the store a standard VHS machine is suitable for a small music-to-picture studio, with certain limitations of course. The only extra equipment necessary in a MIDI sequencer set-up would be a SMPTE-to-MTC converter which can take in standard SMPTE timecode and convert it to MIDI timecode which the sequencer can understand, and any decent sequencer, hardware or software, will understand MTC these days. SMPTE-to-MTC converters are not too expensive, and often a computer's MIDI interface will include such a feature as standard.

The starting point in using a system like this is to ask the production company of the programme for a VHS copy with SMPTE timecode on the sound track. This is standard procedure and not a problem at all. It is also possible to have dialogue on one track and timecode on the other but of course you will need a VHS machine with stereo outputs or you will have a problem! Figure A2.4 – which is very simple but I thought I would include it for the sake of completeness – shows the set-up. One potential problem with a set-up like this is that the SMPTE-to-MTC converter may have difficulty with the quality of timecode coming from the VHS. It very much depends on circumstances. If your VHS is of the so-called hi-fi variety then it has a pair of FM modulated audio tracks recorded on the same area of the tape as the picture. These are of quite reasonable quality and timecode should work fine. Of course, the work copy of the video has to have been recorded on such a machine or the FM tracks won't be there. If there are no FM tracks, or you don't have a hi-fi VHS, then you will be reading timecode from the conventional linear audio tracks which are, as

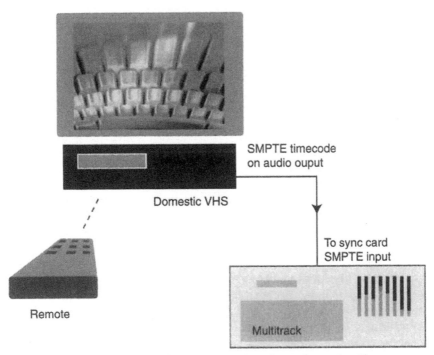

Figure A2.4 A simple code-only master system using a domestic video machine and a digital multi-track with sync card or built-in sync facilities.

you know, just about the lowest quality audio medium around (excepting only half-speed dictation machines) – and how long ago was the audio head cleaned? Whether you get acceptable results depends on the timecode reader in the SMPTE-to-MTC converter. Some will make perfect sense of very poor quality timecode, others will give up at the slightest glitch. The only reliable indicator is to buy from a dealer who will help you out if you have a problem – if there is a moral with timecode and synchronization it is not to cut corners and to do everything in the best possible way.

Assuming that the problems mentioned above either don't happen or are sorted out, a MIDI composer using such a system will achieve perfect satisfaction, and can even improvise music directly into the sequencer while watching the picture as a piano player would in an old silent movie theatre. The next stage is to mix the music onto a synchronizable medium which would form, as far as the composer is concerned, the end product. At this level of the business mixing can take place at the composer's home studio, which is good since a MIDI system is not easily transportable and the data distributed among the many and varied pieces of equipment is notoriously volatile. In fact nothing more than a simple DAT is required even though you couldn't sync it back to the picture at home other than by finger sync. DAT is so stable that timing accuracy can be maintained over even the longest cues and the music can be laid back to the picture in the post-production studio without any problem as long as there have been no mistakes made. If dialogue or sound effects are called for, then greater precision is required and really a timecode DAT would be essential. In this case the DAT could be striped with timecode and the sequencer synchronized to that while the audio is mixed (make sure the start time of the timecode is the same as the video or, if you can, regenerate the timecode from the video and stripe the DAT with that).

'Real' instruments

So far what I have said refers to a MIDI sequencer set-up. Versatile though MIDI may be, there are only two instruments that can be used – the synthesizer and the sampler. MIDI controllable guitars, violins and saxophones have not, for some unaccountable reason, been invented yet. For the composer working in a home studio who wants to use 'real' instruments a step up in technology is necessary. A few years ago this would have meant tape, but synchronizing tape, as we shall see, poses significant problems. Fortunately our modern-day composer can, at relatively low cost, buy a hard disk multi-track recorder that will synchronize to timecode. Hard disk recorders are as efficient as MIDI sequencers and wherever you wind the video, the sequencer and the hard disk recorder will follow, and commence playback within a fraction of a second.

Hard disk recorders have many advantages for multi-track recording but they do have a few drawbacks too, such as the incompatibility between the wide variety of systems on sale, the problem of back-up if removable media are not used, and the cost of the media if they are removable! Having weighed up the pros and cons, many studios still see tape – digital tape of course since analogue is now finished as far as post-production is concerned – as the preferred solution. Our composer, who has had it easy so far but is now going to start struggling under the weight of technology, may be in a situation where tape is seen as essential. The problem is inherent in any tape medium: to get to a certain point in the programme you have to wind the tape, and this takes time. Imagine the scenario now . . . the composer has reached the end of the half-hour programme he is working on and wants to go back to the beginning. He presses rewind on the video's remote. As soon as the video drops out of play, timecode ceases to be issued. As soon as this happens the audio machine has no option but to enter stop mode since all it knows is that there isn't any timecode and maybe the composer is just taking a break. The video winds all the way to the start of the programme, and starts playing and issuing timecode once more. The audio machine realizes that there is now timecode present again, but at a point almost half an hour earlier in time. It starts to rewind and makes its best effort at hitting the moving target with which it is now presented. It misses of course and inevitably has to struggle to catch up. This process is lengthy and surprisingly irritating when it happens a hundred times during the course of a session. There has to be a better way.

CTL and direction

You need nothing more than a cheap domestic video to run a synchronized system with a code-only master. Pay a little more and you will get a video which can output CTL pulses and direction information as well as timecode. CTL or control pulses are a normal thing in video and are recorded along the edge to direct the player to where the video information is to be found. A useful by-product is the fact that they can be used roughly to synchronize two video machines, or audio to video. Working with a video machine with this facility is much faster. When the video is stopped then the audio machine will stop too. When the video rewinds, it will issue CTL pulses at a fast rate and also indicate that the tape is going in the reverse direction. The result is that the audio recorder can wind as the video winds and both will arrive at the same spot on the tape ready to start playing in the minimum time, depending on the sophistication of the synchronizer employed. If the audio machine can wind faster than the video machine, then hopefully the synchronizer will recognize this and slow down the wind (there is a trick to this) so that it keeps pace with the video, otherwise when the video is put into play then the audio machine will have gone some way past the correct point on the tape. If the video winds faster than the audio machine then we still have the problem of hitting a moving target mentioned earlier. Note that our composer is still pressing the buttons on the video, so the tricks that he can get up to are still rather limited, but a music composer's needs are not as complex as those of a dialogue or sound effect editor, for whom there are more suitable alternatives. The dedicated post-production studio too will require more centralized control rather than operating the video machine directly which will, with the right equipment, largely cure the problems mentioned so far.

Burnt-in timecode

The key to being able to extract precise 'hit points' from the video and complement them musically or otherwise is to be able to identify each individual frame and know precisely when it occurs. Of course each frame is uniquely identified with SMPTE timecode on the audio track so this requirement seems to be met. Unfortunately, longitudinal timecode (LTC), as ordinary timecode is often known, can only be read while the tape is running. The output from the tape drops to zero when it is stationary, and all the timecode reader can do is guess at the correct value. So you may be looking at a crystal-clear still image of smoke emerging from the barrel of a gun, but the timecode reader might be guessing a frame or two either way when it actually occurs. The solution is to make a copy of the video and run the video signal through a character inserter which inserts characters corresponding to the timecode from the audio track, which as the tape is moving should be accurate. The result is a tape with LTC on the audio track (regenerated of course) and burnt-in timecode too (Figure A2.5). By the way, burnt-in timecode is often confused with VITC, which is something else entirely.

VITC

Timecode as used on an audio tape or on the audio track of a video tape can also be called longitudinal timecode or LTC. Another type of timecode which only applies to video recorders is VITC or vertical interval timecode, pronounced 'vitsee'. In every frame of video there are a number of unused lines which are a hangover from an earlier technological era. These lines form a useful repository for data, as is seen in closed captioning where text can be encoded into a broadcast TV signal for people with suitably equipped receivers to view. As the video signal is recorded onto tape, VITC is encoded into this part of the

Figure A2.5 In-picture timecode display created by a VITC reader and character inserter. Burnt-in timecode looks similar.

signal, identifying frames in the same way as LTC. The difference is that even when the video is still framed, the video head is still traversing the tape and the VITC can be read as easily as the picture. The advantage is that where LTC is useless in still frame mode and the LTC reader is only guessing at the correct value, the VITC reader knows exactly which frame is being viewed. As an extra facility, VITC can be used to generate characters which can be inserted into the picture to view on screen in a similar manner to burnt-in timecode, the advantage being that if necessary the characters can usually be moved around the screen to avoid obscuring an important part of the action. VITC and burnt-in timecode are easily confused so it is important to be aware of the difference.

Synchronizer systems

A composer can work quite happily in a converted garage or basement with a standard domestic video machine and a MIDI system with a SMPTE-to-MTC interface. Or if he or she wants to record instruments other than synthesizers and samplers, then a hard disk with the appropriate sync option can synchronize very quickly to the video, wherever the video's remote control commands it should go. Tape, even digital tape, is more of a problem since it takes time to wind from one end of the tape to the other, whereas a hard disk recorder is instantaneous. With a simple code-only master system using a synchronizable multi-track tape recorder you would spend a lot of time rewinding the video, putting it in play, and then waiting for the audio machine to start to catch up. This is not the fault of the recorder, it is just the nature of tape. So if you need to use tape, then a little more sophistication is necessary. Of course you could always ask why should you bother to use tape anyway in the modern age of hard disks? I don't have to answer that, the number of tape-based multi-tracks still in use – and still coming off dealers' shelves – answers the question for me.

The key to swift synchronization with tape is to use a more professional video machine, not necessarily a more professional format – VHS will still be OK since all you need is a picture you can work to (and see people's lips move) and a decent quality sound track for the SMPTE timecode. A professional video machine will in addition to standard functions provide outputs for all the transport commands, and also CTL pulses that flow at a rate proportional to the speed of the tape in fast wind. The synchronizer, now rather more than just a card installed within the machine or a timecode-aware remote control, therefore not only knows which way the tape is winding, it knows exactly how fast, and can get the audio machine to the right spot on the tape as quickly as possible. But the next step up is a system which offers a central controller so that, in appearance at least, the controller itself is the master and every other machine is equally subservient to it. In reality, in systems other than those few that incorporate a 'virtual master', one machine must provide the source of timecode and all other machines will be slaves to it. With a system such as this come all the add-ons and tweaks that will make synchronizing tape-based systems almost as smooth as disk systems can be, but of course there will always be the rewind time to consider.

System extras

What should you expect a well-specified synchronizer system to provide that you might not get from a basic sync card? The first is that it should incorporate a timecode generator, although stand-alone generators are also available. Timecode generators vary from very basic to very sophisticated. Basic timecode generators can be set for frame rate (probably not all the possible types) and start time. Do this, press the start button and you're striping tape. This might seem to be enough, but very often it isn't. First of all, with a simple timecode generator such as this, you cannot stripe video tape with timecode. Well you can, but once you are finished it won't work properly. A video recording, as you know, has a frame rate of 25 fps in Europe. If there are 25 frames of video in a second, then there must be 25 frames of timecode in a second too. Each frame of timecode must precisely match a single frame of video and never drift. Therefore the timecode generator must have a video input so that when you stripe, the video frames and timecode frames do indeed match exactly. If they don't, it is a bit like taking a 35 mm stills camera film strip in for a reprint, and you can't decide what the number of the negative you want printed is because there are two numbers on the edge of the film within the width of the frame. This happens because your camera has no way of synchronizing to the numbers which are pre-exposed during manufacture. It is a very similar situation.

Another similar situation is digital audio. If a digital audio recorder is to be striped with timecode, then in each second there must be precisely 44 100 or 48 000 samples and 25, 29.97 or 30 frames of timecode. In fact when a digital audio recorder is synchronized to a 29.97 fps video copy of a film original, the speed of the video is 0.1% slower than the film due to the fact that it is easier to extrapolate 24 frames of film to 30 frames of video and then run it a little slower than it would be to make 29.97 frames of video out of those original 24 film frames. This implies a digital audio sample rate of 44.056 kHz rather than 44.1 kHz, and a similar adjustment if the higher sampling rate is used. This is known as 'pulldown'. A good timecode generator should have an input for a digital audio signal or clock and be able to generate timecode in exact synchronization. Correctly synchronizing a video recorder, digital audio recorder and timecode generator is vital otherwise you don't really know what might happen. You may get a problem straight away, or you may do a significant amount of work and pass an even bigger problem on to someone else.

Jam on it

Another trick of the well-specified timecode generator is 'jam sync'. This is where the timecode generator reads timecode and generates exactly the same code. 'Why?' you ask. The answer is that whenever timecode is copied it must be regenerated so that it has nice sharp edges and makes perfect sense. Timecode from an analogue track on a video recorder will be very rounded and will almost certainly have drop-outs which the timecode reader will have to interpolate. These kinds of errors must never be propagated through the system or project. Another function of jam sync is to repair 'holes' in a timecode track, or to convert discontinuous timecode that does not run in the regular pattern of hours:minutes: seconds:frames into continuous code. It is also common for timecode generators to generate backwards timecode, and even jam sync to code running in the reverse direction. One use for this is to extend a timecode stripe at the beginning of a programme if insufficient pre-roll time has been allowed. Disappointingly, for perfectly understandable reasons no-one has been able to get this to work with either video or digital tape recorders so you might think it is now just a relic of our analogue past. Film people, however, who inhabit a parallel but different universe, can work with audio running backwards just as easily as when it goes in the right direction so any decent hard disk recorder will incorporate this feature.

Timecode readers sometimes come as stand-alone devices where they are used as the front end of a synchronizer system, or they may be incorporated into the synchronizer itself. As you might expect, timecode readers come in all varieties from those which must have perfect timecode otherwise they just won't work properly, to those which can extract time values from badly corrupted code, and handle dropouts or glitches with relative ease. When a dropout occurs on an analogue track, such as the sound track of a video, then the timecode reader will carry on for a user-settable length of time until good code returns. One slight drawback of this is that in some synchronizer configurations it can be inconvenient when you have stopped the tape but the timecode reader carries on, thinking it is just a big drop-out!

Synchronizers

The essence of the synchronizer is to bring two or more machines to the same point in the programme and keep them running at the same speed. One machine will be designated the master and the others will be slaves. Wherever the master goes, the slaves will follow, or chase, to use the correct terminology. In practical terms there is more to it than this, particularly where tape is involved. One common problem is that the audio timecode and video timecode may be at the same frame rate, but the timecode values don't match. For instance the first scene on the video may start at 01:00:00:00 and the first scene on the audio at 10:00:00:00. I have chosen simple numbers which actually are commonly used as starting values, but they could have been any values within the entire twenty-four hour timecode 'day'. In this case there is an offset of nine hours between the two codes. Offsets occur so frequently that virtually any synchronizer would have the facility to deal with the problem. Either an offset value can be entered by hand, or the two machines can be brought roughly into sync manually and then the offset captured. Either way, it will have to be trimmed to precisely the correct value. It is common for synchronizers to offer the option of trimming an offset quickly, which gets the job done but will have audible effects, or slowly so that a slight discrepancy could be adjusted on-air without alerting viewers.

Often, the timecode on a tape may be discontinuous, perhaps because of editing or perhaps because time-of-day code was used during shooting so there is a gap every time the recorder was stopped. There are a variety of solutions to this, but one is offered by synchronizers which have different types of lock. Frame lock compares the timecode numbers on the slave with the timecode numbers on the master and takes whatever action necessary to bring them in line. But if the timecode on the master is discontinuous then with frame lock active, when the master comes upon a discontinuity the timecode value will jump suddenly and the slave will immediately go into fast wind or rewind to catch up. Auto lock (terminology may differ among manufacturers) handles this by first of all searching for the correct timecode values, then looking only at frame edges. The third type of lock is phase lock which ignores the timecode values completely and just looks at the rate the frame edges go by. Together with these there are also hard lock and soft lock. Hard lock maintains precise synchronization with the master but risks propagating wow and flutter, perhaps in the worst case from a VHS master to a digital multi-track! Soft lock will even out any irregularities.

Over the years, manufacturers have developed synchronizers to a point where the best can not only be interfaced to just about any synchronizable transport, they can also learn the characteristics of individual, perhaps almost worn out, examples to get the best from them. The simplest example, which applies to autolocators too, is where a tape is fast wound to a predetermined point. Ideally the tape would be wound at its fastest possible speed until the last moment when the brakes are applied fully and the tape comes to a halt at precisely the right point. Race car drivers know all about this. If that point is missed, then the tape either remains in the wrong place or the transport has to be nudged into the correct position which takes time, and even mere seconds wasted add up during the course of a session, and used to be the primary cause of irritation in the studio before computers were invented! Modern synchronizers will analyse the performance of the transport, for which they acquire all the necessary information during normal operation, and continually fine-

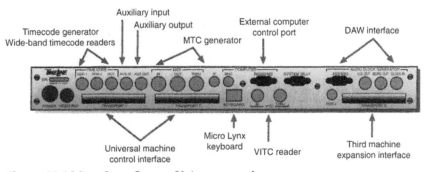

Figure A2.6 Micro Lynx System Unit rear panel.

Figure A2.7 TimeLine Micro Lynx System Unit and Keyboard.

tune their parameters so that each machine can perform at its best. Another aspect of this is in pre-roll, which is the run-up time given before an edit point so all the machines have time to lock to the master. Some synchronizers will individually adjust the pre-roll time of each slave so that lock is achieved in the minimum time possible.

Why 29.97?

It is an enduring mystery why the frame rate for NTSC TV is 29.97 fps rather than 30 fps. How might our lives have been different had 30 fps continued after the transition from monochrome to colour? Of course it was done for a reason, and it really was an amazing triumph to be able to add a colour signal to the existing standard without affecting compatibility with monochrome receivers, other than making adjustment of the (in those days external) horizontal and vertical hold controls necessary when switching between monochrome and colour broadcasts.

The I and Q chrominance signals occupy significant bandwidths of 1.3 MHz and 0.4 MHz respectively. Where did they find room for them? The answer is that the monochrome luminance signal, although it occupies a 4.2 MHz total bandwidth, is full of holes into which the chrominance information can be slotted. Specifically, the energy of the luminance signal is contained in bands that are spaced apart by the 15 750 Hz (at 30 fps) line rate. In between, there is virtually clear space. The chrominance signals were therefore modulated onto a subcarrier of 3.583 125 MHz which is 227.5 times the line rate, chosen by careful calculation so that the energy bands of the chrominance signal, which are similarly spaced, would precisely slot in with those of the luminance signal with only minimal interference between the two. So far so good, but unfortunately the colour subcarrier now interfered with the sound carrier causing a beat frequency with visible effects on the picture. The easy way to solve this would have been to change the frequency of the sound carrier but unfortunately this defeated the whole object of making colour transmissions compatible with existing monochrome receivers. The only alternative was to reduce the frame rate to the familiar 29.97 fps, consequently the line rate reduced to 15 734.263 74 Hz and the colour subcarrier to 3.579 545 MHz. The sound carrier remained the same at 4.5 MHz, the interference was solved, and a whole new era of synchronization problems began!

Synchronization terminology

Address   A timecode value in hours:minutes:seconds:frames.

Burnt-in timecode   Timecode inserted into a video recording which appears on the monitor as part of the picture.

Centre track timecode   Analogue stereo machines may have an extra central track on which timecode can be recorded.

Character inserter   A VITC reader can process the timecode written into the vertical interval of the video signal for use by the synchronizer. Also, a character inserter can be used to superimpose the timecode numbers upon the video picture. This signal can be recorded onto another tape when it becomes burnt-in timecode.

Chase   Where a slave machine follows the movements of a master machine in response to timecode and transport tallies. System control is via the master machine or its autolocator.

Controller   A synchronizer controller allows commands to be sent to all machines in the system, either individually or collectively. Machines can be online or offline.

CTL   Pulses recorded on a video tape proportional to the tape speed in play, fast wind and rewind.

Drop frame   System used to compensate for the non-integer frame rate of 29.97 fps where certain frame numbers are omitted.

ESbus   Joint SMPTE/EBU standard for machine control systems.

Event   A relay closure activated at a specified timecode value. Used for triggering playback on cart machines and CD players which cannot be timecode synchronized.

House sync   A synchronization reference distributed throughout a studio complex.

Interface   Synchronizers need to know the transport commands/tallies and tach characteristics of each machine they are to control. This information is held in an interface. A modern synchronizer has an internal ROM-based directory of interfaces for popular machines.

Jam sync   A timecode generator can lock to an existing timecode source and generate identical numbers. This can be used to extend timecode recorded on a tape, or to re-record a missing section of code.

Layback   When a completed soundtrack is recorded onto the video master.

Lock   When two machines are running synchronously with reference to timecode, they are said to be locked.

LTC   Longitudinal timecode, as recorded on audio recorders or on an audio track of a video recorder.

Master   In a chase system, the engineer operates the controls of the master machine and the synchronizer controls the slave to match the movements of the master. In most larger systems the engineer operates from a synchronizer controller but the timecode reference is taken from the master machine.

Offset   Two tapes which are intended to run synchronously may not have timecode values which correspond. A timecode offset is entered to match the two timecodes.

Online   In a synchronized system, the synchronizer will only control machines which have been designated as online. Other machines are offline.

Play to park   When machines are fast wound to the correct position, the timing reference is taken from tach pulses rather than timecode. These tach references may be a few frames out so if play-to-park mode is used the machines come out of fast wind into play a little earlier than the correct position and play up to the exact spot.

Regeneration   When a timecode reader or reader/generator reads existing code, it reshapes the pulses and corrects any errors. Should always be used when copying timecode.

Slave   Any machine in a synchronized system not designated as the master.

Sony 9-pin (P2 protocol)   A common interface standard.

Striping   Recording timecode onto one track of the tape.

Tally   An output from an audio or video machine from which the synchronizer can determine which transport mode it is in.

Timecode generator   Device for generating timecode and performing regeneration and jam sync.

User bits   Information such as date and reel number, etc. can be inserted into the timecode signal in the form of user bits.

VITC   Vertical Interval Time Code written into gaps in the video signal. VITC can be read in still frame mode, LTC cannot.

Audio in video editing

For most of us now, audio is simply a part of the greater multi-media whole. Only in the fields of music for CD, vinyl and radio broadcast can sound people remain comfortably isolated from video and its strange ways. In post-production, audio has long been considered a poor relation of video, which is rather odd since the power of audio is such that radio is, and always will be, a fully viable medium of information and entertainment – there never has and never will be any form of broadcasting that involves images with no audio content (and I certainly don't count the graphics and text of most Internet websites as broadcasting). Still, we audio types find ourselves in a situation where we have to press continually to be given something approaching equal consideration in everything from budget to studio space to equal access to the coffee machine. To make matters worse, video people commonly deal with audio as part of their everyday working lives and, without wishing to insult those who are managing a remarkable quality of audio considering how much else they have to think about, some of the audio that we hear alongside the images is of a pretty basic standard. Historically, part of the reason for this has been the lamentable quality of the sound tracks of analogue video formats, particularly U-Matic and VHS. A generation of video editors cut their teeth on formats with sound quality so poor it would disgrace a ten-year old car cassette unit that had never had its heads cleaned. And since TV sets before the arrival of digital sound never had anything more than a tiny, barely adequate single internal loudspeaker, the poor quality of the sound didn't matter. Not exactly a recipe for success. Another factor weighing against decent sound in the analogue video era was the fact that then, as now, it is common for video editing to extend to several generations of copying. Oddly enough, on professional equipment the picture can stand it much better than the audio can, but when did you ever hear anyone say, 'I don't think we should risk another generation – the sound can't stand it!'

Fortunately, modern digital video formats are universally provided with adequate audio facilities, so the quality achievable on the soundtrack is no longer an issue. But just because a high standard is achievable does not mean that it is always attained. Indeed, the greater clarity of digital audio means that problems that once would have been masked are exposed for all to hear. Which leads me to the big question: should video editors edit audio along with the picture, or should audio always be handled by specialists? Among audio people of course the answer is that audio should always be handled by audio specialists – that's our bread and butter and the more we can persuade the powers that be that this should be standard procedure, not a luxury, then the more work there will be for us all. But between ourselves I think we have to recognize that this is not always going to be the case. Here are three scenarios:

- In a particular production it has always been the intention that audio should be done separately by specialists. In addition, it is intended that most or all of the original dialogue track will be used. Fairly obviously, it makes no sense for offline video editors – who always get first bite – to ignore the audio. It is digitized from the original tapes along with the video, and whenever a video edit is made it is no trouble at all to cut the audio in the same place. Hence, without any extra effort or initiative, video editors are, almost by default, editing audio. And since they have edited the audio, then what they have done ought to make a good starting point for later work.
- In many situations, time is very much of the essence. Material comes in and has to be assembled into a finished, if not polished, product within the shortest time frame possible. There simply isn't time to go through the process of offline editing the video, which as outlined above automatically entails rough editing the audio, and then sending the work onto another department. Particularly when autoconforming is required, it just takes too long.
- Suppose you are a brilliant video editor. You really have the knack of assembling moving images together in a way that tells the story in the best way it possibly could be told. Since everyone wants you to work for them, it makes sense to buy your own equipment and set up on your own in your attic or garden shed. Where might a sound editor fit into this plan? In another shed? In the cellar? Nowhere, I think.

So there are situations where video editors have to work with audio. I have no disrespect for their capabilities, but it has to be said that when 80% of your attention is on the images, the remaining 20% that is devoted to audio simply cannot allow an equivalent standard of work to what can be achieved by someone who is prepared to devote a full 100% of their attention, and has devoted 100% of their career to date. In an ideal world, a video editor would realize that their 20% commitment can only achieve so much, and would limit themselves to doing things that are safe, useful, and don't create more problems for the people downstream. Or if they are a one-man band, then they would recognize that if they are going to do work of real quality then they will have to put more time and energy into the audio, and perhaps put a little Post-It reminder on the side of their monitor to remind them of that fact.

The editing process

Consider the flow of work in a typical scenario – a television series is shot on Digital Betacam, or DigiBeta to give it its now familiar name. The reels go to the offline suite to be edited on a nonlinear system. Most nonlinear editing systems are not capable of broadcast quality therefore the output of the suite is an EDL which goes with the source reels to the online suite where a finished picture edit is put together onto tape. The materials progress down the line to the audio suite where the segments of audio from the source reels that are actually needed for the finished master are autoconformed into the editing system. Autoconforming follows the picture EDL so that only audio that is linked to picture that is actually used is transferred, plus 'handles' to allow a standard edit to be transformed into a split edit or crossfade if desired, where additional audio is required from either side of the join.

In a scenario such as this, then the video editor doesn't actually touch the audio data – the EDL is just a description of what the editor thinks should be done with it. All the audio is perfectly intact on the source reels, and whatever happens (apart from the tape undergoing spontaneous combustion) it can always be retrieved. In a scenario such as this there is little to go wrong and every opportunity to put things right if necessary, but in practical terms everything has to happen under one roof for this to be the case.

Another scenario might occur when a piece of work is edited purely on DigiBeta, with no offline stage, and the tapes are sent by courier to another part of town. Now, there is no recourse to the source reels except in an emergency. In the real world what will happen is that since video editors are run ragged at the pace they are often asked to work at, audio will get short shrift and there will be bits missing, one half of a stereo track may be blank, dialogue may hop from one audio track to another, material might be mixed together onto one track that shouldn't have been mixed. Now what do we do? In many cases, 'Just make the best of it' is the answer.

Not every audio editor will recognize this scenario, and may even compliment video editors on the high standard of their work. If this is the case then the reason for it is that there is a dialogue between the audio and video people, and mutual respect for their working conditions and particular requirements. It is amazing how many problems can be solved by simply talking to people, but if these problems are not so easily solved elsewhere there is obviously an issue of working methodology to be addressed. One obvious candidate for blame is the complexity of the process, and the variety of routes audio can take in its progression from shoot to screen. Here are a few points for consideration:

- The division between offline and online editing is cumbersome and opens the doors to confusion and misinterpretation of people's intentions. Is there still any need for an offline video editing process?
- Audio is edited in offline video editing suites. But there is no technical or budgetary requirement for audio ever to have to go through an offline process. Can the audio offline stage be eliminated?
- Tape and disk are two very different media. Is it right that they should continue to coexist during the production process?
- Conforming can be a very messy business even when automated – can it be eliminated?

The end of offline?

From the above, we can see that eliminating the offline editing stage would be a positive step. The only reason the offline/online process exists is that, historically, online editing suites have contained massively expensive equipment – exceeding the cost of even high end audio gear. It was not viable to spend time making decisions in the online suite, therefore the decision-making process was carried out in a cheaper offline suite. Once the edit is finished, all those decisions lie in the EDL which can be taken on a floppy disk to the online suite and the editing carried out with maximum efficiency. But what happens when offline equipment starts to challenge online in terms of quality? Surely something has got to give way? We are actually at a point now where equipment that has so far been seen as offline has achieved picture quality close or equivalent to online. I am thinking particularly of nonlinear editing systems which have gradually crept towards broadcast image quality and in some cases are now absolutely there, being able to work to a video data reduction ratio of 2:1, the same as Digital Betacam.

Once this level of quality has been achieved, a number of other things follow. One difficulty in the offline/online process has been the handling of transitions and effects. (A transition is the changeover from one shot to the next, a cut being the simplest.) Although there are a number of standard transitions available which can be specified offline and easily replicated online, if you want to do anything fancy then the best you can do is conceive of the idea offline and spend expensive time in the online suite executing it. But once an offline system has grown to the point where it achieves online quality, suddenly it becomes practical to stuff the system with effects of all kinds, including character generation and everything you might need to finish a programme. Onlining just isn't necessary any more.

So what implications does this have for sound? Let us look at the work flow: shooting will still be on DigiBeta, or other tape format. Although there are such things as disk recording camera formats, tape still rules and will probably continue to do so for some time yet. So the tapes come in with pictures, dialogue, and perhaps a few sound effects and ambiences. They are digitized into a nonlinear editing system at broadcast quality resolution and edited. The sound is rough edited along with the picture. At this point, image-wise the quality is broadcastable, but the data resides on disk rather than tape. Yes, you can broadcast from disk, but outside of news you probably don't want to, disk formats being as diverse as they are. So the work is transferred to tape and sent to the transmission suite, or to audio for further sweetening. Oh dear, we are getting those old problems again. Not all of the sound is on the tape – there aren't any handles for split edits or crossfades – tape can't handle handles! So the audio editor has to go back to the source reels and it's back to that old routine of autoconforming and sitting by a hungry VCR feeding it tapes.

So what is the answer? What will make the flow of work as smooth and straightforward as ideally it should be? The answer is to make the video and audio editing systems compatible. If a project can be video edited with a rough audio edit, and then transferred on disk to the audio suite, then things suddenly become a lot easier. It has to be said that disks are cumbersome and expensive compared with tape at the moment, but in this situation their other advantages outweigh any problems. Since a disk can carry additional information in terms of audio (and video) handles, and can have many more tracks than the measly four of DigiBeta, audio is not compromised in any way, and even if the video editor has done something silly with the sound, there should be additional data on the disk to correct the error. If the worst came to the worst, then the box full of source reels could be brought out again, but this should not be necessary. They can be held in reserve. One way of doing this is to vertically integrate picture and sound. Certain manufacturers make video nonlinear editing systems and specialist audio nonlinear systems which have compatible file formats. The disks too are fully compatible and no conforming is necessary, and tape is only used at the very beginning of the project for shooting, and at the end for transmission and archiving. A very sensible solution, and a network can eliminate the hassle of running the disk down the corridor.

A further benefit can be realized if video and audio editing are performed on a standard desktop computer. Other specialized applications can run alongside the main application to achieve additional flexibility. Another significant plus is that all the networking capability that comes as standard with a personal computer can be employed to advantage in video and audio editing. Sound effects and library music

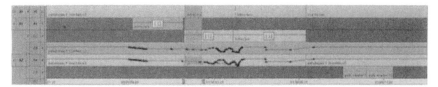

Figure A3.1 Timeline of a nonlinear video editing system (Avid Media Composer).

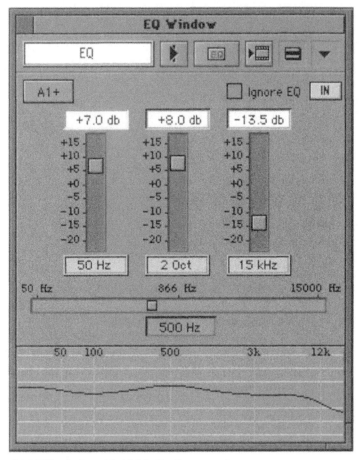

Figure A3.2 EQ window (Avid Media Composer).

can be stored on a central server at very low cost in terms of hardware and learning curve. Add Internet access and you can download material from online sources such as production music libraries and it is there ready for use on the desktop. Try that with your dedicated audio workstation.

In conclusion, it seems that it is indeed perfectly possible to do high quality audio work in a video environment, which means that in our

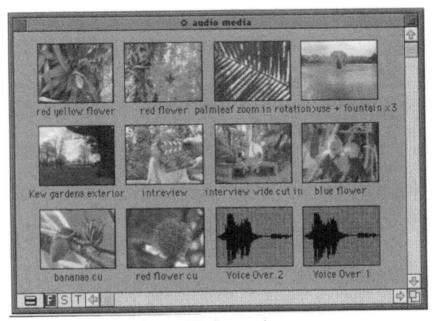

Figure A3.3 Bin window (Avid Media Composer).

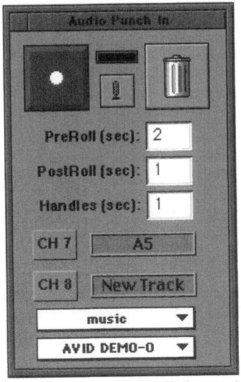

Figure A3.4 Audio punch-in window (Avid Media Composer).

audio enclaves we are going to have to take additional pains to do that work better or more efficiently. But I have only told half of the story. If video editors can now edit audio, how long will it be before audio editors can edit video? The answer to that is that it is perfectly feasible, if the manufacturers are prepared to give us the tools to do it. Whether we would want to acquire the story-telling skills of a video editor is another matter, but as the market for visual product expands and fragments then there must be opportunities available for the bold. Anyway, who hasn't at one time or other thought that a video cut could have benefited from modification, when viewed in the context of the finished audio edit. So far it hasn't really been feasible, but there may come a time when everyone is working online, and offline doesn't exist any more. Then, who knows?

Index

2:3 pulldown, 77, 210–11
3D film, 158, 199
8-VSB system, 142–4
16mm film, 161, 163–4
35mm film, 148, 149, 159, 161, 163, 164
65mm film, 156–7, 161, 163, 164, 192–200

ACATS *see* Advisory Committee on Advanced Television Service
additive colour mixing, 246
advanced television (ATV), 138–40
Advanced Television Standards Committee (ATSC), 139–40, 142
Advisory Committee on Advanced Television Service (ACATS), 139
Alchemist Ph.C standards converter, 79
aliasing, 74–5
ALiS (Alternate Lighting of Surfaces), 92
AM *see* amplitude modulation
Ampex:
    digital video recording, 55–6
    video recording, 7–8, 19–23, 25
amplitude modulation (AM), 144–5
analogue-to-digital conversion, 35–9
anamorphosis, 155–6, 157, 161, 162, 189
ATSC *see* Advanced Television Standards Committee (ATSC)
ATV (advanced television), 138–40

audio:
    see also sound
    Betacam, 46, 47–9
    crosstalk, 24–6
    flat screen displays, 95
    home cinema, 96, 97, 99–104
    PAL speed-up, 214
    pulldown, 212
    timecodes, 257–8
    video editing, 273–80
    video recording, 21, 24–6, 260–2
auto-tint, 17
azimuth recording, 25, 26, 56

B picture *see* bi-directional picture
back porch, 14
Baird, John Logie, 1, 3, 4, 10, 202
band ripple, 56
bandwidth:
    chrominance signals, 15–16
    digital television, 122–3, 133
    television, 12–13
    video recording, 18–19
Barnack, Oskar, 163
BBC, 3–4
beam indexing, 84–5
Betacam, 45–6
    see also Digital Betacam
Betacam SP, 46–9, 65–7, 68–9
Betamax, 24, 26, 45–6
bi-directional (B) picture, 124–6
black body locus, 247, 249
black halo, 37

black and white film, 164–5
Block Cross Interleaved
    Reed-Solomon code, 49
Blondes, 222
bluescreen, 234–41
Bosch, 22
Broads, 222

C-Format, 8–9, 22–4, 46, 66
C-Reality, 205–6, 207, 208
cable digital television, 134
camcorders, 45–6, 61–2, 65–72
camera tube, 4
cameras:
    see also camcorders
    colour, 33–4
    damage, 31
    film, 44–5
    IMAX, 193–5
    scanning, 10, 28
    television, 4, 27–43
captions, 111
carbon arc lights, 186–7, 218–19,
    221
cathode ray tube, 1, 28–31, 84–8,
    93–4, 202–5
CBS, 5–6
CCD *see* charge-coupled devices
CCITT (International Telegraph and
    Telephone Consultative
    Committee), 114, 122
charge-coupled devices (CCD), 32–3,
    34–9, 204
    Betacam, 47
    component video, 46, 58
    motion problems, 77
    sampling, 75
Chrétien, Henri, 155
chroma key, 238–9
chrominance signals, 15–17, 58
    Betacam, 46–7
    C-Format, 66
    digital video recording, 52
    DV, 60
    DVCPRO, 63–4
    eye sensitivity, 115
    phase integrity, 55
CIE *see* Commission Internationale de
    l'Eclairage

CIE colour triangle, 247–8
cinema technology, 183–91
CinemaScope, 155–6
Cinerama, 154–5, 192
Cintel, 202, 204, 205, 206, 207
Coded Orthogonal Frequency
    Division Multiplex (COFDM),
    133–4, 142
coercivity, 178–9
COFDM *see* Coded Orthogonal
    Frequency Division Multiplex
colorimetry, 38
colour, 243–50
    see also chrominance signals
    Betacam, 46–7
    cathode ray tube, 84–5
    CCD cameras, 38–9
    compute monitors, 113–14
    definition, 14
    equation of, 15
    eye sensitivity, 115
    film, 157–8, 170–1, 173–4, 176
    sync pulse, 82
    telecine, 206–8
    television, 4–6, 14–17
    video recording, 18–19, 21
colour burst, 16, 17, 19, 21, 55
colour grading, 170–1, 173–4, 207
colour temperature, 220–2, 224, 248–9
colour triangle, 247–8
colour under, 24
comet tailing, 31
Commission Internationale de
    l'Eclairage (CIE), 247
component signals, 58
    analogue television systems, 73–4
    Betacam SP, 49
    digital video recording, 52–3
    video, 46–7, 66
composite signals, 46, 58
    analogue, 55
    C-Format, 66
    digital video recording, 52–3, 55
    standards conversion, 75–6
    sync, 82–3
compositing, 234–41
compressed time division multiplexed
    system, 46–7
compression, 113–31
contact printing, 172

contrast range, 31, 39, 201–2, 206, 218–19
control (CTL) pulses, 21, 263–5, 271
control pulse, 21
convergence, 84
Crookes, William, 1
cross-colour, 37
crossed curves, 176
crosstalk, 24–6, 58, 257
CTL *see* control

D1, 51–5, 67–8
D2, 55–6
D3, 56–7
D5, 57
DAB *see* Digital Audio Broadcasting
dailies *see* rushes
dark current, 31, 34–5
DAT, 26, 262
DCT *see* discrete cosine transform
De Forest, Lee, 1
detail correction, 36–7
dichroic filters, 33–4, 225–7
Dickson, K.L., 147–8
diffraction grating, 250
digital audio, Betacam, 48–9
Digital Audio Broadcasting (DAB), 137
Digital Betacam, 65–72, 275
Digital Cinematography, 63–6, 70–1
Digital Light Processor (DLP), 94
digital signal processing (DSP), 35–9
digital television (DTV), 122–3, 130, 132–45
digital video, 51–72
    compression, 132–3
    pulldown, 212–13
    sampling, 74
    timebase correction, 23–4
Digital Video Broadcasting (DVB), 133–4, 135
digitization, 107
discrete cosine transform (DCT), 60, 115–17, 124, 129–30
dissolve, 109–10
DLP *see* Digital Light Processor
DMX 512, 232–3
Dolby, Ray, 7, 19, 48, 152
Dolby C, 48

Dolby Digital, 142
Dolby Pro-Logic, 97, 98, 99, 100, 102
Dolby SR, 101
Dolby Stereo, 101–2
Dolby Surround, 96, 98, 100, 101–3
Dolby Virtual Surround, 103, 104
domestic video formats, 24–6
DSP *see* digital signal processing
DTV *see* digital television
DV format, 59–61
DVB *see* Digital Video Broadcasting
DVB Project, 133
DVCPRO, 59, 61–4
DVD Video, 123, 130
DVTR, 212–13

Eastman, George, 147–8, 163
Edison, Thomas Alva, 147–8, 163
edit decision list (EDL), 107, 173, 275
editing, 105–12
    film, 105–6, 107, 109, 173
    nonlinear, 60, 62–3, 67, 105–12, 173
    timecode, 251–72
    video, 251–72, 273–80
EDL *see* edit decision list
electromagnetic radiation, 243–4
Electronic News Gathering (ENG), 44–5
EMI, 3, 202
ENG *see* Electronic News Gathering
entropy coding, 117–18
equation of colour, 15
error correction, 53–4
    DTV, 143
    DV, 60
    DVB, 134
Exposure Index (IE), 165–6
exposure latitude, 165
eye, 244–5
    black to white transition, 114–15
    colour, 14–16
    persistence of vision, 10, 147
    selectiveness of, 132
    sensitivity for chrominance, 115
    spatial frequencies, 114, 115, 123–4

Farnsworth, Philo T., 10, 30
FED *see* Field Emission Display

Federal Communications Commission
     (FCC), 4, 5, 6, 139–41, 142
Field Emission Display (FED), 93
field-sequential system, 5–6
fields, 12–13
film, 63–6, 146–62
  black and white, 164–5
  bluescreen, 240
  cameras, 44–5
  colour, 157–8
  contrast range, 201–2, 206
  Digital Cinematography, 70–1
  editing, 105–6, 107, 109, 173
  history, 146–53
  IMAX, 192–200
  laboratories, 167–8, 174–5
  lighting, 218–21
  motion problems, 77–8
  printing, 171–3
  processing, 167–76, 199–200
  projection, 183–91
  pulldown, 209–17
  sampling, 74
  sound, 258–60
  stock, 163–7
  telecine, 77, 138, 166, 201–8, 209–17
finger sync, 260
fixed pattern noise, 32–3, 34–5
flat panel displays, 89–95
flicker:
  film, 10, 149, 183, 185
  television, 5, 210
flying spot telecine, 202–6
FM *see* frequency modulation
forced development, 176
Fourier analysis, 115
Frame Interline Transfer CCD, 35
frame rate, 16
  film, 10, 149, 150, 152, 156–7, 183
  IMAX, 193, 196–7
  NTSC television, 209, 211–13, 270–1
  pulldown, 209–17
  television, 11–12
  timecode, 354–6
frame storage area, 35
frames, 12
France, 6, 148
frequency modulation (FM):
  Betacam, 48
  U-Matic, 24

video recording, 19, 21
fresnel lens, 222–4
front porch, 14

gain, telecine, 206
gain boost, 35–6, 42
gamma control, 206
Geneva movement, 183–4
geometry, 90–1
'ghost' signals, 142–3
Ginsberg, Charles, 7, 19
glass shots, 234–5
gobos, 229–31
GOPs *see* Groups of Pictures
Grand Alliance, 140, 142
Groups of Pictures (GOPs), 124–6
guard bands, 24–6

HAD *see* Hole Accumulated Diode
halation, 167
HDTV *see* High Definition Television
helical scanning, 8–9, 21–4, 60
Hi-8 camcorders, 26, 62
High Definition Television (HDTV),
     93, 130, 134
  film format, 164
  scanning, 142
  standards, 74
  USA, 138–9, 141–2
HMI (hydragyrum medium arc length
     iodide) lamps, 221–2
Hole Accumulated Diode (HAD), 34
home cinema, 96–104
hotheads, 227
hue, 226, 248
Huffman code, 118
Hyper HAD, 34–5

I pictures *see* intraframe pictures
iconoscope, 4, 29–30, 202
illuminance, 250
image compression, 113–31
image orthicon, 30–1
IMAX, 157, 161, 183, 192–200
interlaced scanning, 12, 142
interleaving, 55, 210
intermediate film, 167

International Standards Organization
(ISO), 114, 121
interpolation, 76–7
intraframe (I) pictures, 124–6
ISO *see* International Standards
Organization

jam sync, 267–8, 272
Japan, HDTV, 139
jitter, 19, 21, 22, 23, 83
Joint Photographic Experts Group
(JPEG), 114
JPEG image compression, 60, 106, 113,
114–21, 124

KeyKode logging, 173

lamphouse, 186–7
lamps, 219–21
Laptop Field Editing System, 62
LaserDisc, 98–9
Lauste, Eugene Augustin, 150, 151,
152
LCD *see* Liquid Crystal Displays
Lee Filters, 226
lenses:
  IMAX, 194
  projectors, 189
LEP *see* Light-Emitting Polymer
levels (MPEG2), 128–30, 131
lift, 206
light, 14, 243–4
light valve technology, 18
Light-Emitting Polymer (LEP), 94
lighting, 218–33
  CCD cameras, 32
  contrast, 39
  tube cameras, 31
linearity, video recording, 18, 19
lines (television), 3–4, 12–13
Liquid Crystal Displays (LCDs), 84,
  89–91, 92
longitudinal timecode (LTC), 264, 272
LTC *see* longitudinal timecode
Lucasfilm, 99–100
luma key, 236–8
Lumière, Auguste and Louis, 148–9

luminance signal, 15–16, 58
  Betacam, 46–7
  C-Format, 66
  colour under, 24
  digital video recording, 52
  DV, 60
  DVCPRO, 63–4
  image compression, 115
luminous flux, 250
luminous intensity, 250

macroblocks, 126–7
magnetic sound track, 153, 155, 258–9
magnetic tape, 177–82
Marconi, 3
Martin Professional, 231
Master Set-up Unit (MSU), 39–43
matte, 235–40
matte line, 239
MIDI timecode, 260–2
moiré, 58
monitors, 81–95
monochrome waveform, 13–14, 15
motion compensation, 77–80
motion control photography, 240–1
motion estimation prediction, 126–7
Motion Picture Experts Group
(MPEG), 60, 114
MPEG2 image compression, 122–31,
  133, 142–4
MPEG image compression, 113, 114,
  121
MSU *see* Master Set-up Unit
multi-path reception, 142–3
MUSE (Multiple Sub-Nyquist
  Sampling Encoding), 139
Muybridge, Eadweard, 147

National Television Standards
  Committee (NTSC), 4, 6, 209
negative breakdown, 169
Neopilot, 212, 258–9
Nipkow disc, 2–3, 4–5
Nipkow, Paul, 2–3
noise:
  Betacam, 47–8
  CCD cameras, 32–3, 34
  compression, 67–8

noise – *continued*
 detail enhancement, 37
 gain boost, 36
 Hyper HAD, 35
 optical sound, 152
nonlinear editing, 60, 62–3, 67,
 105–12, 173
NTSC system, 6, 73–4
 2:3 pulldown, 77, 210–11
 colour, 15–16, 17
 composite form, 58
 digital signal processing, 37–8
 digital video recording, 52, 55
 frame rate, 209, 211–13, 270–1
 Laserdisc, 98–9
 pulldown, 77, 209–13
 standards conversion, 65, 66, 75–7
 vestigial sideband transmission,
 144–5

OCL *see* On-Chip Lens technology
offline editing, 107, 110–11, 275–7
Olsen, Harry F., 7
Omnimax, 199
On-Chip Lens technology (OCL), 35
online editing, 107, 110–11, 276–7
optical printing, 172
optical sound, 150, 151, 152–3,
 189–90, 258

P pictures *see* predictive pictures
PAL 1200, 231
PAL (Phase Alternate Line) system, 6,
 73–4
 colour, 16, 17
 composite form, 58
 digital video recording, 52, 55, 56
 pulldown, 209, 213–16
 standards conversion, 65, 66, 75–7
Panasonic, 56–7, 59, 61
performance lighting, 224–9
persistence of vision, 10, 147
Phase Alternate Line *see* PAL
photomosaic, 29–30
Pilotone, 212, 258–9
Pioneer, 98–9
pixels, 90–1, 92, 93, 94, 113, 126
plasma displays, 91–3, 97

platters, 187–9
Plumbicon tube, 35
pre-knee, 35–6
prediction error, 126–7
predictive (P) pictures, 124–6
printing, film, 171–3
profiles (MPEG2), 128–30, 131
progressive scanning, 11, 87, 92, 142
projection, 183–91, 195–7
projection televisions, 18, 84, 97
projectors, 183–90
pulldown, 77, 209–17
pushing, 176

QPSK *see* Quadrature Phase Shift
 Keying
Quadrature Phase Shift Keying
 (QPSK), 134
Quadruplex system, 7–9, 19–22
quantization, 124, 129–30

radio, DAB, 137
raster, 11, 29, 82
RCA, 4, 5–6, 30
 camcorders, 46
 LCD, 89
 optical sound, 152
 shadow mask, 85
 video recorders, 7–9
 video recording, 18, 19, 21
receivers, 28, 81–2
Redheads, 222
Reed-Solomon coding, 49, 143, 144
reels of film, 187–9
Reeves, Hazard, 155
release print, 174–5
Remote Control Panels, 43
retentivity, 178
reversal film, 165
RGB signal, 58
Rocobscan, 231
Rolling Loop, 196–7
rotoscoping, 235–6, 237
rushes, 168–71, 199–200

S-VHS, 26, 97
safe areas, 160, 162, 163

sampling:
   pulldown, 212–13
   rates, 36
   standards conversion, 74–5, 76–7
Sarnoff, David, 4, 18
satellite digital television, 134
saturation, 226, 248
scanning, 2–4
   cameras, 10, 28
   DTV, 142
   HDTV, 142
   helical, 8–9, 21–4, 60
   interlaced, 12, 142
   monitors, 83
   progressive, 11, 87, 92, 142
   telecine, 205–6
   television, 11–13
   transverse, 7, 20
   video, 7, 8–9, 20, 21–4
SECAM (Sequential Colour with
      Memory) system, 6, 73, 74
segmentation, 56
separation positive prints, 176
Sequential Colour with Memory see
      SECAM
set-top boxes, 137
shading, CCD cameras, 39
shadow mask, 85–6, 87, 94
shuttle mode, 55, 56, 62
silver enhancement, 176
Skin Tone Detail, 69
skin tone detail enhancement, 37, 38
sloping verticals, 77
SMPE, 149
SMPTE timecode, 75, 251, 253, 260–2,
      264
Society of Motion Picture and
      Television Engineers (SMPTE), 8,
      251
   colorimetry, 38
   timecode, 251, 253, 260–2, 264
   video formats, 22
Sony:
   Betacam, 45, 48, 49
   cameras, 40–1
   cathode ray tube, 88
   D1, 51–2, 54
   Digital Betacam, 68
   digital video recording, 51–2, 54,
      56, 59

helical scan video recording, 22
Hyper HAD, 34–5
video, 8–9
sound:
   see also audio
   film, 45, 150–3, 258–60
   film projection, 189–90
   IMAX, 199
   synchronization, 258–72
sound advance, 152–3, 190
spatial compression, 123–4
spatial frequencies, 114, 115, 123–4
standard illuminants, 47, 249
standards conversion, 73–80, 209, 215,
      216
Steadicam, 195
stepped diagonals, 37
stereoscopic images, 158
Strand Lighting, 222
subtractive colour mixing, 246
Super 16, 161, 162, 163–4
Super 35, 156
surround sound, 96, 97, 98, 99
synchronization, 82–3, 144, 258–72

Technicolor, 157–8
telecine, 138, 201–8
   film stock, 166
   motion problems, 77
   pulldown, 209–17
television, 10–17
   see also NTSC: SECAM; PAL
   ATV, 138–40
   cameras, 4, 27–43
   colour, 4–6, 14–17
   digital, 122–3, 130, 132–45
   frame rate, 11–12, 209, 211–13,
      270–1
   HDTV, 74, 93, 130, 134, 138–9,
      141–2, 164
   history, 1–6
   projection, 18, 84, 97
   receivers, 28, 81–2
   safe areas, 162, 163
   standards, 209
   telecine, 77, 138, 166, 201–8, 209–17
temporal compression, 124–6
terrestrial digital television, 133–4,
      135

THX Home Cinema, 99–101
timebase circuit, 83
timebase correction, 23–4, 47, 51, 55
timecode, 212–13, 216, 251–72
timeline, 108–10
timing stability, 18–19
Todd, Mike, 156–7
Todd-AO, 156–7
transverse scanning, 7, 20
Trellis encoder, 143–4
Trinitron cathode ray tube, 88
triode vacuum tube, 1
tube cameras, 28–31, 33–4, 77, 79

U-Matic, 24, 45, 46, 107, 274
Ultimatte, 239–40
Ultra Panavision, 157
United Kingdom:
   digital television, 132–8
   television, 3–4
United States of America:
   digital television, 138–45
   pulldown, 209
   television, 4, 209
   timecode, 254–6

Vari-Lite, 225–31
VARI*LITEs, 225–31
versioning, 111–12
vertical hold, 82
vertical interval timecode (VITC),
   264–5, 272
vertical smear, 32, 35
vestigial sideband transmission, 144–5
VHS recorders, 26, 97–8

video:
   see also digital video
   analogue-to-digital conversion,
      35–9
   audio, 260–2, 273–80
   contrast range, 201–2
   editing, 105–6, 107, 112, 251–72
   history, 7–9
   lighting, 218–21
   monitors, 81–95
   recording, 18–26
   standards conversion, 209
   synchronization, 260–2
   timecode, 251–72, 264–5
video cameras, sampling, 74–5
video compression, 67
Video on Demand (VOD), 131, 136
video mapping, 53
video tape, 177–82
Video-8, 26
vidicon tube, 4
VistaVision, 156
Vitaphone system, 150–2
VITC *see* vertical interval timecode
VOD *see* Video on Demand

'wagon wheel' effect, 74–5
Waller, Fred, 154
white, 247
white balance, 35–6
widescreen:
   film, 153, 154–6
   television, 138

Zworykin, Vladimir, 4, 29, 202